AF213302

Semidefinite Programming in Quantum Information Science

Online at: https://doi.org/10.1088/978-0-7503-3343-6

IOP Series in Quantum Technology

Series Editor: **Barry Garraway** (School of Mathematical and Physical Sciences, University of Sussex, UK), **Barry Sanders** (Institute for Quantum Science and Technology, University of Calgary, Canada), and **Lincoln Carr** (Quantum Engineering Program, Colorado School of Mines, USA)

About the series

The IOP Series in Quantum Technology is dedicated to bringing together the most up to date texts and reference books from across the emerging field of quantum science and its technological applications. Prepared by leading experts, the series is intended for graduate students and researchers either already working in or intending to enter the field. The series seeks (but is not restricted to) publications in the following topics:

- Quantum biology
- Quantum communication
- Quantum computation
- Quantum control
- Quantum cryptography
- Quantum engineering
- Quantum machine learning and intelligence
- Quantum materials
- Quantum metrology
- Quantum optics
- Quantum sensing
- Quantum simulation
- Quantum software, algorithms and code
- Quantum thermodynamics
- Hybrid quantum systems

A full list of titles published in this series can be found here: https://iopscience.iop.org/bookListInfo/iop-series-in-quantum-technology

Semidefinite Programming in Quantum Information Science

Paul Skrzypczyk
School of Physics, University of Bristol, Bristol, UK

Daniel Cavalcanti
Algorithmiq Ltd, Helsinki, Finland

IOP Publishing, Bristol, UK

© IOP Publishing Ltd 2023

All rights reserved. No part of this publication may be reproduced, stored in a retrieval system or transmitted in any form or by any means, electronic, mechanical, photocopying, recording or otherwise, without the prior permission of the publisher, or as expressly permitted by law or under terms agreed with the appropriate rights organization. Multiple copying is permitted in accordance with the terms of licences issued by the Copyright Licensing Agency, the Copyright Clearance Centre and other reproduction rights organizations.

Permission to make use of IOP Publishing content other than as set out above may be sought at permissions@ioppublishing.org.

Paul Skrzypczyk and Daniel Cavalcanti have asserted their right to be identified as the authors of this work in accordance with sections 77 and 78 of the Copyright, Designs and Patents Act 1988.

ISBN 978-0-7503-3343-6 (ebook)
ISBN 978-0-7503-3341-2 (print)
ISBN 978-0-7503-3344-3 (myPrint)
ISBN 978-0-7503-3342-9 (mobi)

DOI 10.1088/978-0-7503-3343-6

Version: 20230301

IOP ebooks

British Library Cataloguing-in-Publication Data: A catalogue record for this book is available from the British Library.

Published by IOP Publishing, wholly owned by The Institute of Physics, London

IOP Publishing, No.2 The Distillery, Glassfields, Avon Street, Bristol, BS2 0GR, UK

US Office: IOP Publishing, Inc., 190 North Independence Mall West, Suite 601, Philadelphia, PA 19106, USA

Supplementary material is available at https://github.com/paulskrzypczyk/SDPBook.

Contents

Part II Semidefinite programming in quantum information science

Preface

We can safely say that learning semidefinite programming was a turning point in both of our careers. It all started in 2013, when we both joined Prof. Antonio Acín's group at ICFO—The Institute of Photonic Sciences (Barcelona) to conduct our postdoctoral research. At that time we became interested in the problem of characterising quantum steering, a form of quantum correlations that were first noticed by Schrödinger in response to Einstein, Podolsky and Rosen's famous 1935 paper. We suspected that there was more to discover about quantum steering, and the role it can play in quantum information science. It was then, during the traditional Quantum Information Conference '13, held in the beautiful mountain village of Benasque (Spain), that Miguel Navascués (now a group leader at the University of Vienna) mentioned that steering could be studied using semidefinite programming. We immediately started digging into this idea and, a few months later, we published our first paper (also in collaboration with Miguel) showing how to use semidefinite programming in the quantification of steering. Since then, we have both been using semidefinite programming directly to achieve results in quantum information and foundations, or indirectly to test assumptions, to the point that using this technique has become an obsession. In fact, this is the main reason why we decided to write this book: semidefinite programming has helped us tremendously, and we are sure that it will help other researchers too.

We also have to mention the push we got from the VII Paraty Quantum Information School in 2019, where one of us was invited to give a mini-course on the topic of this book. During this event we saw a big interest from students on the topic, and realised there was a lack of a concise and coherent source of information about it, that could appeal to theorist or experimentalist, mathematician or physicist, and all those in between.

Because of this, we would like to thank Antonio Acín for providing us with a relaxed and inspiring collaborative environment in his group. We would also like to thank Miguel Navascués, and everyone involved in the organisation of the Benasque Quantum Information Conference as well as the organisers and students of the VII Paraty Quantum Information School. We also thank our numerous colleagues who, for so many times, have helped us to better understand semidefinite programming. In particular Marco T Quintino, Dennis Rosset, Jean D Bancal, Valerio Scarani, Marco Piani, Ashley Montanaro, Tony Short, and Sandu Popescu. Finally, we would like to thank IOP Publishing, and in particular John Navas, for giving us the opportunity to write this book.

Author biographies

Paul Skrzypczyk

Paul Skrzypczyk is currently an Associate Professor and Royal Society University Research Fellow at the University of Bristol in the School of Physics. He obtained his PhD in Theoretical Physics from the University of Bristol in 2011, under the supervision of Professor Sandu Popescu, with his PhD studies focusing on quantum nonlocality and quantum thermodynamics. He carried out postdoctoral research at the University of Cambridge, and ICFO—The Institute for Photonic Sciences, before returning to Bristol in 2015. In 2016 he was awarded a Royal Society University Research Fellowship, and became a lecturer in 2018 and an Associate Professor in 2022. Paul's research interest span many areas of quantum information and quantum foundations, ranging from quantum nonlocal effects, such as Bell nonlocality, quantum steering and quantum teleportation, to quantum measurements and measurement incompatibility, to quantum thermodynamics. Convex geometry, semidefinite programming and convex optimisation have been the primary mathematical tools used in his research over the years.

Daniel Cavalcanti

Daniel Cavalcanti is currently a senior researcher at Algorithmiq Ltd. He obtained a PhD in Theoretical Physics at ICFO—Institute of Photonic Sciences in 2008 on the topics of entanglement and characterisation of quantum correlations. After a short postdoc at ICFO in 2009, he joined the Centre for Quantum Technology (Singapore) in 2010, first as a postdoc in Professor Valerio Scarani's group, and soon as an independent researcher. In 2013 he returned to ICFO, where he stayed until 2021 on a Ramón y Cajal grant. Daniel's research has focused on quantum foundations, quantum correlations, quantum communication and, more recently, quantum computation. Daniel also holds a master's degree in graphic design and runs Bitflow, a graphic design studio dedicated to science and technology related projects.

About this book

Semidefinite programming (SDP[1]) is a type of optimisation problem with vast applications in physics, engineering, combinatorial optimisation, and many other fields. In this book we are interested in applications of semidefinite programming in the field of quantum information. It turns out to be pretty natural that many problems in quantum mechanics and quantum information can be cast as SDPs, because the mathematical description of quantum states and quantum measurements in terms of positive semidefinite operators fit naturally in the SDP framework. Thus, although here we will mostly focus on quantum information, the present book can also be relevant when addressing other problems in quantum physics.

From a practical point of view, there are two main reasons to study SDPs. First, once a problem is recognised as an SDP, there is a theoretical machinery that can be used to solve the problem or, at least, obtain bounds on its solution. Second, there are plenty of efficient computer algorithms and modelling languages for solving SDPs, such as CVX, CVXOPT, CVXPY, MOSEK and YALMIP, allowing us to numerically solve SDPs involving relatively big matrices with today's standard laptop computers. As a starting point for performing numerical calculations of SDPs we recommend in particular the CVX and CVXPY modelling systems for constructing and solving convex optimisation problems (of which SDPs are a subclass) that can be implemented in Matlab and Python respectively. This book will focus exclusively on the theoretical machinery of SDPs. However, to accompany it, we have set up a repository at https://github.com/paulskrzypczyk/SDPBook, which contains a small number of example codes, covering some of the SDPs studied in this book.

The structure of this book is the following. It is split into two parts. Part I is devoted to the description of semidefinite programming, focusing on its main aspects. It starts with a simpler subclass of optimisation problems called *linear programs* (chapter 1) and then moves on to the main description of SDPs (chapter 2). In these two chapters we will also set up the notation, nomenclature and basic definitions and concepts that will be used throughout the book. Then, Part II is devoted to particular problems from across the realm of quantum information science that can be tackled with the help of semidefinite programming. The list of problems we consider here is by no means exhaustive. Nonetheless, it is important to explain a little about the choice of problems presented. First, we wanted to cover a wide range of problems in quantum information. In this regard, you will find chapters related to quantum states, measurements, entanglement and channels. Second, we have chosen problems specifically that allow us to introduce key methods and tools for transforming problems into SDPs, and for manipulating and analysing those SDPs. These methods and tools are those which we believe to be very useful for a reader who wants semidefinite programming to be a useful tool in

[1] Throughout this book we will use the acronym SDP to denote both *semidefinite programming* and *semidefinite program*.

their research or work. In this respect, we encourage the reader to go through all chapters, even if the particular topic of the chapter is not so relevant for them; the way the topic is studied should hopefully teach them something new and interesting about semidefinite programming beyond the topic itself. In any case, we list the most important messages of each chapter in a section called *Concluding Remarks.*

The list of problems we cover are as follows:

- Chapter 3 addresses one of most basic problems in quantum physics, how to determine the properties of quantum states produced by an uncharacterised source. There are numerous variants of this problem, the most well-known being quantum state tomography, that is, estimating exactly which state the source is emitting. In this chapter we will review this and other variants of state estimation, including the so-called quantum marginal problem. This will allow us to study various aspects of SDPs, including how to obtain certificates of infeasibility and the prevalent notion of a 'witness'.
- In chapter 4, we shift focus to quantum measurements. We will first discuss the problem of estimating which measurement a given measuring device is performing. We then study quantum state discrimination, the problem of determining—by performing a measurement—which state, out of a set of possibilities, a given source is producing. We show how to gain new insight into this problem by using the key concept of SDP duality, and how the tool of complementary slackness can be used to derive optimality conditions.
- We then move on to the problem of characterising quantum entanglement in chapter 5. Entanglement is nowadays seen as the main resource behind many quantum information tasks, such as quantum teleportation, computation and cryptography. In this chapter we will see how SDPs can be used to detect and quantify entanglement. We will also see how we can use a sequence—or hierarchy—of semidefinite programs to obtain approximations to the set of separable quantum states, and therefore bounds on many quantities of interest.
- In chapter 6 we will study one of the most basic and interesting properties of quantum mechanics: measurement incompatibility. We will see how this problem can be solved by semidefinite programming, and use duality to uncover an unexpected link to quantum nonlocality.
- Finally, in chapter 7 we will study quantum channels. We will see how the so-called Choi–Jamiołkowski isomorphism—which links quantum channels with quantum states—allows for semidefinite programming techniques to be applied to quantum channels. We will then see how we can estimate an uncharacterised channel, calculate the diamond norm (which characterises the operational distinguishability between channels) and finally how to use duality to uncover a link between a channel optimisation problem and quantum entropies.

At the end of each chapter we also suggest a small amount of additional material on each topic (in the section *Further reading*). It is important to note that the list of references is by no means exhaustive. Our aim is simply to provide the curious reader with further resources—in the form of relevant textbooks or review articles—to complement the present text. In this connection, let us stress that there are several texts on semidefinite programming that can be used to complement this book in general. In particular, the book *Convex Optimization* by S Boyd and L Vandenberghe is a must-have reference on convex optimisation, which includes and goes beyond the theory of SDPs, and provides the reader with a comprehensive amount of information on the subject (and is freely available!). More related to quantum information are the textbook *The Theory of Quantum Information* and lecture notes *Theory of Quantum Information* by J Watrous and the lecture notes *Semidefinite Programming & Quantum Information* by J Sikora and A Varvitsiotis, all of which are freely available online.

Finally, this book is very much just an introduction to the topic of semidefinite programming. There are many more advanced topics that we chose not to cover here. Our goal was to provide a solid foundation to the key aspects of the theory, and to demonstrate their widespread applicability in quantum information science. We hope that using this book as a foundation, those readers who choose to do so, will be well placed to go on and master more advanced aspects of semidefinite programming—and more generally convex optimisation—and put them to good use in whichever direction they see fit.

Part I

The fundamentals

IOP Publishing

Paul Skrzypczyk and Daniel Cavalcanti

Chapter 1

Linear programming

1.1 The basics

Before jumping into semidefinite programming (SDP), we start by discussing a simpler type of optimisation problem called linear programming. This will help us in setting up some of the notation, and will lay important foundations for the reader, allowing them to become more familiar with the type of problems that we will consider throughout this book.

A linear program (LP[1]) is a constrained optimisation problem that consists in maximising (or minimising) a *linear* function $f(x_1, \ldots, x_k)$—called the *objective function*—of a set of real variables $x_1, \ldots, x_k$, i.e. $f: \mathbb{R}^k \to \mathbb{R}$. We will use the colour orange for variables throughout the book, to visually highlight that these are the unknowns that need to be solved for.

These variables have to satisfy a set of *linear equality* and/or *inequality* constraints. That is, we have a set of m functions $g_i: \mathbb{R}^k \to \mathbb{R}$ with associated values b_i, for $i = 1, \ldots, m$, such that $g_i(x_1, \ldots, x_k) = b_i$ for all i, and similarly a second set of n functions $h_j: \mathbb{R}^k \to \mathbb{R}$ with associated bounds c_j, for $j = 1, \ldots, n$, such that $h_j(x_1, \ldots, x_k) \leqslant c_j$ for all j.

Combining everything together, we can write a linear program as:

$$\text{maximise} \quad f(x_1, \ldots, x_k) \tag{1.1a}$$

$$\text{subject to} \quad g_1(x_1, \ldots, x_k) = b_1, \quad \vdots \tag{1.1b}$$

$$g_m(x_1, \ldots, x_k) = b_m, \tag{1.1c}$$

$$h_1(x_1, \ldots, x_k) \leqslant c_1, \quad \vdots \tag{1.1d}$$

$$h_n(x_1, \ldots, x_k) \leqslant c_n. \tag{1.1e}$$

[1] Throughout this book we will use the acronym LP to denote both *linear programming* and *linear program.*

doi:10.1088/978-0-7503-3343-6ch1

© IOP Publishing Ltd 2023

The different parts (or elements)—the objective function, the constraint functions, the bounds—that arise in a linear program fall broadly into two different classes; there are those that can be thought of as defining the problem itself, and those that form the particular *instance* of the problem. It is perhaps easiest to think of this in analogy to how one would write a piece of computer code to solve a problem. Instead of rewriting the code every time before being run, it is useful for the code to accept some *input data*, which modifies the problem to some extent. This input data is what specifies the particular instance of the problem, while the code (which doesn't change from instance to instance) specifies the problem that is being solved.

As an example, consider that we want to solve a problem which involves optimising over a probability distribution. In this case, we will need to impose that the probabilities are positive numbers, which sum up to unity. These constraints belong to the problem, and will be the same, independent of the *instance* of the problem being solved. On the other hand, there may be additional constraints on the probabilities that will specify the instance. In example 1.1 below, we additionally constrain some of the marginal probability distributions. Since we are interested not in a specific set of marginals, but rather in the problem for an arbitrary set of fixed marginals, it is natural to view these as the input data, specifying a particular instance of a general marginal problem.

There is no rule regarding which parts or elements of a problem will be fixed by the problem, and which will depend upon the particular instance. For example, it could be that the equality constraints define the problem, or that they define the instance, or that some arise from the problem and some from the instance. In general, each part of a linear program can relate either to the instance or the problem or both, ranging from the objective function, to just the bounds on some of the inequality constraints, etc. As we come to study the many examples of LPs (and SDPs) in this book this should hopefully become clear, and show how rich the interplay between instance and problem can be.

In order to aid our thinking, when writing explicit examples down, we will use the colour blue to denote the input data (i.e. those parts which depend upon the instance) and leave the other elements which define the problem in black.

Example 1.1. Marginal Problem.

Consider three random variables X, Y and Z that can assume values x, y and z respectively. Suppose that we have knowledge of the pairwise marginal distributions $P_{X,Y}(x, y)$, $P_{X,Z}(x, z)$, and $P_{Y,Z}(y, z)$, but not of the joint distribution $P_{X,Y,Z}(x, y, z)$. Consider that we would like to find the maximum of a linear function over all possible joint distributions that have the fixed marginal distributions that we observe. For example, this could be simply one of the elements, say $P_{X,Y,Z}(0, 0, 0)$, or the correlator $\langle XYZ\rangle = \sum_{x,y,z} xyzP_{X,Y,Z}(x, y, z)$. Given that probabilities are positive numbers that sum up to unity, this is an instance of an LP, and can be written as:

$$\text{maximise} \quad f(P_{X,Y,Z}(x, y, z)) \tag{1.2a}$$

$$\text{subject to} \quad \sum_z P_{X,Y,Z}(x, y, z) = P_{X,Y}(x, y) \;\; \forall \;\; x, y, \tag{1.2b}$$

$$\sum_y P_{X,Y,Z}(x, y, z) = P_{X,Z}(x, z) \;\; \forall \;\; x, z, \tag{1.2c}$$

$$\sum_x P_{X,Y,Z}(x, y, z) = P_{Y,Z}(y, z) \;\; \forall \;\; y, z, \tag{1.2d}$$

$$P_{X,Y,Z}(x, y, z) \geqslant 0 \;\; \forall \;\; x, y, z. \tag{1.2e}$$

In this example it is only the right-hand side of the equality constraints, enforcing that the marginal distributions are correct, that are the input data specifying a given instance of the problem.

It is also important to note that there appears to be a missing constraint, namely $\sum_{x,y,z} P_{X,Y,Z}(x, y, z) = 1$. This constraint is in fact implicitly contained in the above problem. If for example we consider the first constraint, and sum both sides over all x and y, as long as $P_{X,Y}(x, y)$ is a normalised marginal probability distribution, which will always be the case in problems of interest, then it implies that the joint distribution must also be normalised.

The above is a simple example, but hopefully starts to demonstrate that natural problems indeed have the structure of the abstract linear program specified above. As will be seen throughout this book, the structure of a linear program is sufficiently general that many problems can be identified as as such.

Exercises

1.1 Show that any LP can also be written as a minimisation problem, by suitably redefining the objective function. *In this way, it can be seen that we can equally consider minimisation or maximisation problems, and we should use whichever is most natural from the perspective of the problem of interest*[2].

1.2 Show that a constraint imposing a lower bound can always be expressed as a constraint imposing an upper bound. *In this way, it can be seen that we can take all constraints to be upper bounds without loss of generality.*

[2] We will state the *take-home message* of an exercise, where appropriate, directly after it, in purple italic text, or provide additional context/information about the exercise and its significance. It will be useful to consult these take-home messages once the exercise has been completed, to make sure the purpose of the exercise is fully understood.

It can often be useful to adopt a vectorial notation and rewrite LPs in a more compact form. First, it is useful to introduce a vector $\vec{x} \in \mathbb{R}^k$ to contain the variables of the problem. Second, the linear function f can be encoded in a vector $\vec{a} \in \mathbb{R}^k$, such that $f(x_1, \ldots, x_k) = \vec{a} \cdot \vec{x}$. In a similar way, each linear function associated with the equality and inequality constraints, g_i and h_j respectively, can be encoded in vectors $\vec{r}_i \in \mathbb{R}^k$ and $\vec{s}_j \in \mathbb{R}^k$ respectively. With this convention, the LP (1.1) can be re-expressed as

$$\text{maximise} \quad \vec{a} \cdot \vec{x} \tag{1.3a}$$

$$\text{subject to} \quad \vec{r}_i \cdot \vec{x} = b_i, \qquad i = 1, \ldots, m \tag{1.3b}$$

$$\vec{s}_j \cdot \vec{x} \leqslant c_j, \qquad j = 1, \ldots, n. \tag{1.3c}$$

As a second example, we now show how the ℓ_1 and ℓ_∞ norms of vectors can be cast as LPs. This will show that in some cases, even if the objective function is not linear, it is possible to manipulate the optimisation problem so that it can be written as an LP. This is an important point that will appear again and again in other problems discussed in this book. This shows that LPs capture a much broader class of problems than may initially be obvious.

Example 1.2. ℓ_1 and ℓ_∞ norms.

For a vector $\vec{y}$, recall that the ℓ_1 and ℓ_∞ norms are given by

$$\|\vec{y}\|_1 = \sum_i |y_i|, \qquad \|\vec{y}\|_\infty = \max_i |y_i|. \tag{1.4}$$

Although both of these norms appear to be *non-linear* functions of $\vec{y}$, they can both in fact be calculated by linear programming.

In order to see how this is possible, consider first the simpler problem of evaluating the absolute value $|z|$ of a number z. This will be equal to z if $z \geqslant 0$, and equal to $-z$ if $z < 0$. One way to evaluate this is to introduce a variable x, and solve the simple LP

$$|z| = \text{minimise} \quad x \tag{1.5a}$$

$$\text{subject to} \quad -x \leqslant z \leqslant x, \tag{1.5b}$$

where we have expressed *two* inequality constraints in a single line, which should hopefully cause no confusion. If $z \geqslant 0$, then the constraint $z \leqslant x$ will be the limiting one, and $x = z$ is the minimum value of x we can take. On the other hand, if $z < 0$, then $-x \leqslant z$ will be the limiting one, in which case $x = -z$ is the minimum (positive) value of x. In both cases, $x = |z|$, as required.

This basic idea can be generalised to evaluate the ℓ_1 and ℓ_∞ norms In particular, we find

$$\|\vec{y}\|_1 = \text{minimise} \quad \vec{1} \cdot \vec{x} \tag{1.6a}$$

$$\text{subject to} \quad -\vec{x} \leqslant \vec{y} \leqslant \vec{x} \tag{1.6b}$$

where $\vec{1} = (1, \ldots, 1)$, such that the objective function is the sum of the elements of $\vec{x}$, that is $\vec{1} \cdot \vec{x} = \sum_i x_i$. This LP is then seen to just sum up the absolute value of each component of $\vec{y}$, as required. For the ℓ_∞ norm we have, similarly,

$$\|\vec{y}\|_\infty = \text{minimise} \quad x \tag{1.7a}$$

$$\text{subject to} \quad -x\vec{1} \leqslant \vec{y} \leqslant x\vec{1}. \tag{1.7b}$$

Here, instead of introducing a vector variable $\vec{x}$, we have introduced a *scalar* variable x. The constraint demands that $x \geqslant |y_i|$, for all i. It is the largest such value, $\max_i |y_i|$ which will be the limiting constraint, and thus we will recover the ℓ_∞ norm.

Exercises

1.3 If $f(x_1, \ldots, x_k) = \vec{a} \cdot \vec{x}$, show that the component a_k of $\vec{a}$ is $f(0, \ldots, 0, 1, 0, \ldots, 0)$, where the 1 is the kth argument, corresponding to the variable x_k.

1.4 Consider the function of two variables, $f(x_1, x_2) = 2x_1 - x_2$. Write down the associated vector $\vec{a}$.

1.5 If $g_i(x_1, \ldots, x_k) = \vec{r}_i \cdot \vec{x}$, find the analogous relationship between the components of $\vec{r}_i$ and the value of g_i at a carefully chosen point.

1.6 An alternative formulation of the ℓ_1 norm compared to (1.6) is

$$\|\vec{y}\|_1 = \text{maximise} \quad \vec{t} \cdot \vec{y} \tag{1.8a}$$

$$\text{subject to} \quad \|\vec{t}\|_\infty \leqslant 1, \tag{1.8b}$$

i.e. the ℓ_1 norm of a vector $\vec{y}$ is the largest scalar product between $\vec{y}$ and any vector $\vec{t}$ whose ℓ_∞ norm is bounded by one. For this reason, the ℓ_1 norm is said to be the *dual norm* of the ℓ_∞ norm.

The formulation (1.8) is not an LP, due to the non-linear constraint. Using the LP (1.7) for the ℓ_∞ norm, show that it is possible to obtain an alternative LP representation of the ℓ_1 norm given by

$$\|\vec{y}\|_1 = \text{maximise} \quad \vec{t} \cdot \vec{y} \tag{1.9a}$$

$$\text{subject to} \quad -\vec{1} \leqslant \vec{t} \leqslant \vec{1}. \tag{1.9b}$$

It is also common to go one step further than the vectorial form for an LP given in (1.3), by combining all of the vectors $\vec{r}_i$ associated to the equality constraints into a single rectangular $m \times k$ matrix B, such that the rows of B are given by the transposed vectors (row vectors) $\vec{r}_i^{\,\mathsf{T}}$,

$$B = \begin{pmatrix} \vec{r}_1^{\,\mathsf{T}} \\ \vdots \\ \vec{r}_m^{\,\mathsf{T}} \end{pmatrix}. \tag{1.10}$$

Similarly, all of the vectors associated to inequality constraints can be combined into a rectangular $n \times k$ matrix C, such that the rows are given by the transposed vectors $\vec{s}_j^{\mathsf{T}}$,

$$C = \begin{pmatrix} \vec{s}_1^{\mathsf{T}} \\ \vdots \\ \vec{s}_n^{\mathsf{T}} \end{pmatrix}. \tag{1.11}$$

The values b_i and bounds c_j can then naturally be viewed as the components of vectors $\vec{b} \in \mathbb{R}^m$ and $\vec{c} \in \mathbb{R}^n$, respectively, and (1.3) can be expressed as

$$\text{maximise} \quad \vec{a} \cdot \vec{x} \tag{1.12a}$$

$$\text{subject to} \quad B\vec{x} = \vec{b}, \tag{1.12b}$$

$$C\vec{x} \leqslant \vec{c}. \tag{1.12c}$$

It is important at this stage to point out that in (1.12c) we use the notation $\vec{y} \leqslant \vec{z}$ in a component-wise manner, meaning that $y_k \leqslant z_k$ for $k = 1, \ldots, n$. Alternatively, we can view this as imposing that the vector $\vec{z} - \vec{y}$ is component-wise nonnegative.

1.1.1 Feasibility

A vector $\vec{x}$ satisfying the constraints of an LP, e.g. (1.12b) and (1.12c) is called a *feasible point*. In general, there will exist infinitely many feasible points for an LP, as will be seen in more detail below. We will denote the set of all feasible points by $\mathcal{F}$, and refer to it as the *feasible set*. This set has an extremely important property: it is a *convex set*. This means that if $\vec{x}_1$ and $\vec{x}_2$ are both feasible points, then any point of the form $\vec{x}' = p\vec{x}_1 + (1-p)\vec{x}_2$, with $0 \leqslant p \leqslant 1$ is also a feasible point, $\vec{x}' \in \mathcal{F}$. A proof of this is as follows:

Proof

$\vec{x}_1 \in \mathcal{F}$ means that $B\vec{x}_1 = \vec{b}$ and $C\vec{x}_1 \leqslant \vec{c}$. Similarly, $\vec{x}_2 \in \mathcal{F}$ means that $B\vec{x}_2 = \vec{b}$ and $C\vec{x}_2 \leqslant \vec{c}$. Together these conditions imply, due to linearity, that for any p, the point $\vec{x}' = p\vec{x}_1 + (1-p)\,\vec{x}_2$ satisfies

$$\begin{aligned} B\vec{x}' &= B(p\vec{x}_1 + (1-p)\vec{x}_2), \\ &= p\vec{b} + (1-p)\vec{b}, \\ &= \vec{b}, \end{aligned} \tag{1.13}$$

and so $\vec{x}'$ satisfies the necessary equality constraints to be in $\mathcal{F}$. Similarly, if we restrict to $0 \leqslant p \leqslant 1$, then furthermore, again due to linearity,

$$\begin{aligned} C\vec{x}' &= C(p\vec{x}_1 + (1-p)\vec{x}_2), \\ &\leqslant p\vec{c} + (1-p)\vec{c}, \\ &= \vec{c}, \end{aligned} \tag{1.14}$$

where the restriction on p is needed to ensure that in both cases the inequalities do not change direction. Thus $\vec{x}'$ satisfies all the necessary constraints, and is contained in $\mathcal{F}$.

We will discuss the geometrical meaning of convexity in the next section.

With the above notation in place for the feasible set of an LP, we can compactly write an LP as

$$\alpha = \max_{\vec{x} \in \mathcal{F}} \vec{a} \cdot \vec{x}, \tag{1.15}$$

that is, as the maximisation of the objective function over the feasible set.

In the above we have also introduced α to denote the *optimal value* of the problem, i.e. the maximal value that the objective function can take on the feasible set. When no confusion arises, we will sometimes omit the word 'optimal', and refer simply to '*the value*' of an LP, as the optimal value of the objective function. We will use the notation $\vec{x}^*$ to denote an *optimal variable* that achieves the maximum, i.e.

$$\vec{x}^* = \operatorname*{argmax}_{\vec{x} \in \mathcal{F}} \vec{a} \cdot \vec{x}. \tag{1.16}$$

Optimal variables in general are not unique: there can be infinitely many optimal variables that simultaneously achieve the maximum. In fact, the set of all optimal variables is itself a *convex set*. The proof of this is very similar to the proof that the feasible set $\mathcal{F}$ is convex, and is left as an exercise below.

Clearly, the value of the objective function evaluated at any feasible point provides a lower bound on α, i.e.

$$\vec{a} \cdot \vec{x} \leqslant \alpha \quad \forall \vec{x} \in \mathcal{F}. \tag{1.17}$$

It can also occur that there is no feasible point, i.e. that there is no $\vec{x}$ that satisfies all of the constraints of the LP; another way of saying this is that the feasible set is the empty set: $\mathcal{F} = \varnothing$. In this case, the LP is said to be *infeasible*.

It is customary, and as we will see later, useful and mathematically consistent, to assign the optimal value $\alpha = -\infty$ in this case. Intuitively, this indicates that the problem takes the smallest possible value.

This finally leads us to a special subset of LPs known as *feasibility LPs*. Here the objective function is constant, and can without loss of generality be taken to be such that $\vec{a} = \vec{0}$, i.e. such that $f(x_1, \cdots, x_n) = 0$ for all $\vec{x}$. The associated LP is then equivalent to checking for the existence of a feasible point $\vec{x}$. We have that

$$\alpha = \max_{\vec{x} \in \mathcal{F}} \vec{0} \cdot \vec{x} = \begin{cases} 0 & \text{if } \mathcal{F} \neq \varnothing, \\ -\infty & \text{if } \mathcal{F} = \varnothing, \end{cases} \tag{1.18}$$

which is to say that such an LP checks whether the feasible set $\mathcal{F}$ is the empty set or not.

Since it is a rather cumbersome notation to introduce an arbitrary constant objective function, we will use a special notation for feasibility problems, which emphasises that their goal is to find a feasible point if it exists:

$$\text{find} \quad \vec{x} \tag{1.19a}$$

$$\text{subject to} \quad B\vec{x} = \vec{b}, \tag{1.19b}$$

$$C\vec{x} \leqslant \vec{c}. \tag{1.19c}$$

Interestingly, this shows that feasibility problems of the form (1.19) are indeed themselves linear programs, even though they appear not to be of the form (1.12) due the lack of a maximisation.

An example of a feasibility LP that naturally arises is the following:

Example 1.3. Majorisation (feasibility form).
Consider two probability distributions $P(x)$ and $Q(x)$. Majorisation captures the important notion of one probability distribution being more 'disordered' than another. The distribution $P(x)$ majorises $Q(x)$, written $P(x) \succ Q(x)$, if there exists a doubly-stochastic matrix D, with components D_{xy}, such that $\vec{q} = D\vec{p}$, where the components of $\vec{q}$ and $\vec{p}$ are the probabilities $q_x = Q(x)$ and $p_y = P(y)$ respectively. This leads to the following *feasibility LP*:

$$\text{find} \quad D \tag{1.20a}$$

$$\text{subject to} \quad D\vec{p} = \vec{q}, \tag{1.20b}$$

$$\sum_x D_{xy} = 1 \;\; \forall \; y, \tag{1.20c}$$

$$\sum_y D_{xy} = 1 \;\; \forall \; x, \tag{1.20d}$$

$$D_{xy} \geqslant 0 \;\; \forall \; x, y. \tag{1.20e}$$

In the above, the constraints (1.20c) and (1.20e) ensures that the matrix D is stochastic, so that it maps probability distributions to probability distributions (i.e. maintains positivity and normalisation), while (1.20d) ensures that D leaves the uniform distribution invariant, and so is moreover doubly stochastic.

Exercises

1.7 Show that the set of all optimal variables of a linear program forms a convex set. *This implies, in particular, that if two or more distinct optimal variables exist, then there will be infinitely many optimal variables, formed by convex combinations of these. This fact can prove useful, as we will see later.*

1.8 Check if the following LPs are feasible or not:

(a)

$$\begin{aligned} \text{find} \quad & \vec{x} \\ \text{subject to} \quad & x_1 + x_2 \geqslant 1, \\ & x_1 \leqslant \frac{1}{3}, \\ & x_2 \leqslant \frac{1}{2}. \end{aligned}$$

(b)

$$\begin{aligned} \text{find} \quad & \vec{x} \\ \text{subject to} \quad & x_1 + x_2 + x_3 = 1, \\ & x_1 + x_2 \geqslant 1, \\ & x_2 + x_3 \geqslant 1. \end{aligned}$$

1.9 In example 1.1 we saw an instance of the marginal problem, where we want to calculate a function $f(P_{X,Y,Z}(x, y, z))$ of the unknown joint distribution of three random variables. An important instance of the marginal problem is the *feasibility* form, where one wants to know whether the marginals are *compatible* with *any* joint distribution or not.

(a) Write down the feasibility marginal problem LP for the scenario considered in example 1.1, where we are given all three pairwise marginal distributions as data.

(b) Show that the marginal problem is always infeasible if the single party marginals are inconsistent. That is, for example, if

$$\sum_y P_{X,Y}(x, y) \neq \sum_z P_{X,Z}(x, z).$$

This gives a simple set of necessary conditions that must be satisfied—and should always be checked—before the LP is solved.

(c) Consider the marginal problem where we are only given two of the pairwise marginal distributions, $P_{X,Y}(x, y)$ and $P_{Y,Z}(y, z)$, and assume these are consistent, i.e. $\sum_x P_{X,Y}(x, y) = \sum_z P_{Y,Z}(y, z)$. Show that this marginal problem is *always feasible*, with solution

$$P^*_{X,Y,Z}(x, y, z) = \frac{P_{X,Y}(x, y)P_{Y,Z}(y, z)}{P_Y(y)}. \tag{1.21}$$

We say that this marginal problem has no frustration between the constraints—there is no tension between simultaneously satisfying both constraints, and the random variable Y can be simultaneously correlated with both X and Z. In Chapter 3, when we consider the analogous quantum marginal problem, it will be seen that this property no longer holds, due to monogamy of entanglement.

It can sometimes be useful to recast a feasibility LP as a standard LP—i.e. as an LP with a non-constant objective function. There are a number of reasons why this can be useful. First, from a numerical perspective it is often much more stable to solve a standard LP than a feasibility LP. Second, if a problem is indeed infeasible, it can often be useful to understand *how close* it is to being feasible. Finally, as we will see below, when we introduce the important concept of *duality*, having a feasibility LP expressed as a standard LP can again prove useful.

There are multiple ways in which a feasibility LP can be turned into a standard LP, and depending upon the particular problem, different methods might be preferable. The feature they have in common is to *relax* the constraints defining the feasible set, by introducing additional variables, and then to minimise (or sometimes maximise) these variables, to attempt to solve the problem with the original feasible set $\mathcal{F}$. The example below shows one way of doing this for the Majorisation feasibility problem from example 1.3.

Example 1.4. Majorisation (maximisation form).
The uniform probability distribution $U(x) = 1/n$, for $x = 1, \ldots, n$ is majorised by all other probability distributions—it is the most disordered of all probability distributions. Using this fact, we can consider the family of probability distributions

$$Q_t(x) = tQ(x) + (1 - t)U(x) \tag{1.22}$$

parametrised by a new variable t such that $0 \leqslant t \leqslant 1$. This family extrapolates between $Q(x)$ and the uniform distribution $U(x)$. When $t = 0$, we have $P(x) \succ Q_0(x) = U(x)$. We can then look for the largest value of t such that $P(x) \succ Q_t(x)$: if this occurs at $t = 1$, this means $P(x) \succ Q(x)$, and the original LP is feasible. If it occurs at $t < 1$, then $P(x) \not\succ Q(x)$, and the problem is infeasible. Moreover, the closer t is to 1, the closer $P(x)$ is to majorising $Q(x)$. This relaxed form of the problem is an LP, given by

$$\text{maximise} \quad t \tag{1.23a}$$

$$\text{subject to} \quad D\vec{p} = t\vec{q} + (1 - t)\vec{u}, \tag{1.23b}$$

$$\sum_x D_{xy} = 1 \ \forall \ y, \tag{1.23c}$$

$$\sum_y D_{xy} = 1 \ \forall \ x, \tag{1.23d}$$

$$D_{xy} \geqslant 0 \ \forall \ x, y, \tag{1.23e}$$

$$t \leqslant 1, \tag{1.23f}$$

where $\vec{u}$ is the uniform probability vector with components $u_x = 1/n$.

Comparing to the original feasible form (1.20), we see that we have introduced one new scalar variable, which has become the objective function. In terms of the constraints, it is only the first equality constraint which is *relaxed*. Indeed, for a fixed value of t, the right-hand side of this constraint has changed compared to its original form, and can be seen as relaxing the feasible set when $t < 1$.

Exercises

1.10 In example 1.4 above, only the upper bound $t \leqslant 1$ is imposed, and not the lower bound $0 \leqslant t$. Show that the lower bound on t is *irrelevant* due to this being a maximisation problem.
Hint: Show that $t = 0$ is a feasible choice, by finding an explicit choice for D that is simultaneously feasible.

1.11 Turn the two feasibility LPs from exercise 1.8 into standard form LPs. *Hint: Consider relaxing the inequality constraints by introducing a new variable to be minimised.*

1.2 Geometric interpretation

Linear programs can be understood geometrically, and this is in fact a very useful way to gain understanding and intuition about them. In this section, we will outline some of this geometric understanding.

We will first focus on the feasible set $\mathcal{F}$, before considering the objective function. Recall that the feasible set is specified by a collection of equality and inequality constraints. In what follows, we will take the variable to be a point $\vec{x} \in \mathbb{R}^k$.

1.2.1 Equality constraints

The set of all equality constraints $\{g_i(\vec{x}) = b_i\}_i$ specify an *affine subspace* $\mathcal{H} \in \mathbb{R}^k$ in which the point $\vec{x}$ must lie. Here the significance of the subspace being *affine* rather than *linear* is that it need not pass through the origin. The dimension of this subspace will depend upon the number of *linearly independent* equations in the LP. If there are ℓ linearly independent equations then the dimension of $\mathcal{H}$ will be $(k - \ell)$. Each equation $g_i(\vec{x}) = b_i$ itself specifies a *hyperplane*, and the vector $\vec{r}_i$ which appears in the vector representation of the function, $g_i(\vec{x}) = \vec{r}_i \cdot \vec{x}$, is geometrically the *normal vector* to the hyperplane. The affine subspace $\mathcal{H}$ is then nothing but the intersection of all of the hyperplanes specified by each of the equality constraints. An illustrative example involving two constraints in 3-D is depicted in figure 1.1.

1.2.2 Inequality constraints

Inequality constraints, on the other hand, specify *half-spaces*, i.e. they divide the space into two, such that $\vec{x}$ must lie in one half. In particular, the vector $\vec{s}_j$ which appears in the vector representation of the function $\vec{s}_j \cdot \vec{x} = h_j(\vec{x}) \leqslant c_j$ is again geometrically the normal vector to the hyperplane which divides the space in two. This hyperplane is displaced from the origin in the direction of $\vec{s}_j$ by the amount c_j.

The collection of all inequality constraints thus specify a set of half-spaces, and a feasible point must lie in the intersection of all of these half-spaces. The resulting body is called a *polyhedral set*. It may or may not be the case that this polyhedral set is bounded—i.e. is finite in extent. If it is bounded, then the body is called a *polytope*. An illustrative example of a non-bounded polyhedral set formed by two inequality constraints in 2-D is depicted in figure 1.2.

1.2.3 The feasible set

Putting the above two ingredients together, we see that the feasible set $\mathcal{F}$ of an LP is then geometrically the intersection of the affine subspace $\mathcal{H}$ generated by the equality

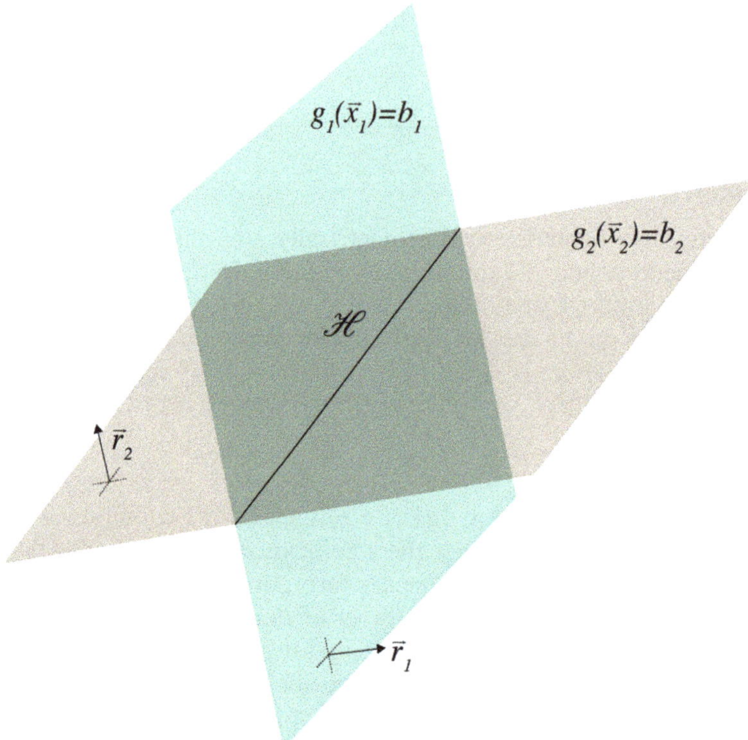

Figure 1.1. *Equality constraints.* An illustrative example in 3-D of how an affine subspace $\mathcal{H}$ arises from two equality constraints, $g_1(\vec{x}) = b_1$ and $g_2(\vec{x}) = b_2$. Both of these constraints specify *planes*, with normal vectors $\vec{r}_1$ and $\vec{r}_2$, respectively. Their intersection is a *line*, depicted in black. Any feasible point $\vec{x}$ of the corresponding LP must lie on this line.

constraints, with the polyhedral set formed by the inequality constraints. This intersection forms itself a *polyhedral set*, which again may or may not be a polytope. We illustrate all of these ideas in the following example, in which we see how the probability simplex arises.

Example 1.5 The probability simplex.
The probability simplex arises as the feasible set of an LP when the variable being optimised is a probability distribution. Consider a three outcome random variable X, which takes on the values $x = 0$, 1 and 2, with associated probability distribution $p(x)$. In order to be a valid probability distribution we have 3 inequality constraints, imposing that the probabilities are positive,

$$p(0) \geqslant 0, \quad p(1) \geqslant 0, \quad p(2) \geqslant 0, \tag{1.24}$$

and a single equality constraint, imposing that the probability distribution is normalised,

$$p(0) + p(1) + p(2) = 1. \tag{1.25}$$

Geometrically, we can represent a probability distribution as a point in $\mathbb{R}^3$. The 3 inequality constraints collectively demand that the point $\vec{p}$, with components $p_x = p(x)$, lies in the *positive orthant*. The equality constraint is the plane which passes though (1, 0, 0), (0, 1, 0) and (0, 0, 1), the three *deterministic* probability distributions. The feasible set this creates is the *probability simplex*, as depicted below.

In this example, the feasible set is seen to be bounded, and so is a *polytope*—it is simply an equilateral triangle.

In higher dimensions we would arrive at a similar construction, which would produce higher dimensional simplices arising as the intersection of the normalisation constraint (geometrically an affine subspace) with the inequality constraints (geometrically the positive orthant), which always leads to a simplex, a polytope with vertices equal to the deterministic probability distributions.

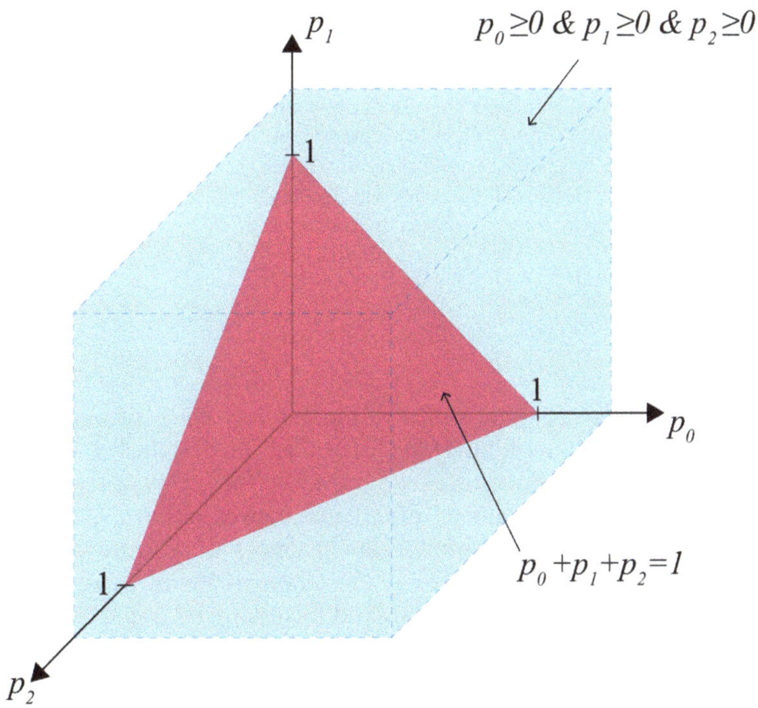

Finally, in section 1.1.1 when we introduced the notion of feasibility, we saw that the feasible set has the extremely important property of being a *convex set*. This is most easily understood from a geometrical perspective. Recall that convexity says

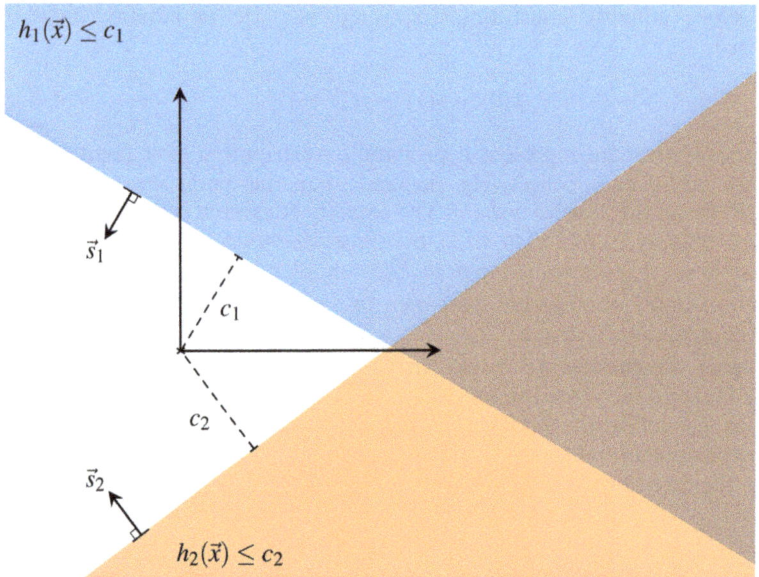

Figure 1.2. *Inequality constraints.* An illustrative example in 2-D of how inequality constraints specify half-spaces. The constraint $h_1(\vec{x}) = c_1$ specifies a line, displaced by a distance c_1 in the direction of $\vec{s}_1$, the vector which defines the function, $h_1(\vec{x}) = \vec{s}_1 \cdot \vec{x}$. Points which satisfy the inequality lie in the (blue) shaded region above this line. It is possible to specify any half-space in this manner. In this example, the brown shaded triangular region on the right contains all the points which jointly satisfy both of the constraints, and is the resulting polyhedral set formed by these two inequality constraints.

that every point of the form $\vec{x}' = p\vec{x}_1 + (1 - p)\vec{x}_2$, with $0 \leqslant p \leqslant 1$ is feasible whenever both $\vec{x}_1$ and $\vec{x}_2$ are. Geometrically says that the *line segment* between any two points inside the set—$\vec{x}_1$ and $\vec{x}_2$—is itself inside the feasible set. That is, the feasible set is not just a polyhedral set or polytope, but moreover a *convex polyhedral set* or *convex polytope*. In figure 1.4 you will find a few examples of different convex sets in 2-D.

Linear programs, and SDPs are special instances of a more general class of optimisation problem known as *convex optimisation problems*. In all such problems, the feasible set always forms a convex set, and this is crucial to the fact that LPs and SDPs don't have *local optima* but only *global optima*, and hence can be solved much more easily than problems which do not have this property (so-called *non-convex optimisation problems*).

1.2.4 The objective function

Having visualised the feasible set $\mathcal{F}$ geometrically, all that remains is to understand the objective function. Recall that this is a linear function $f(\vec{x})$, which can be

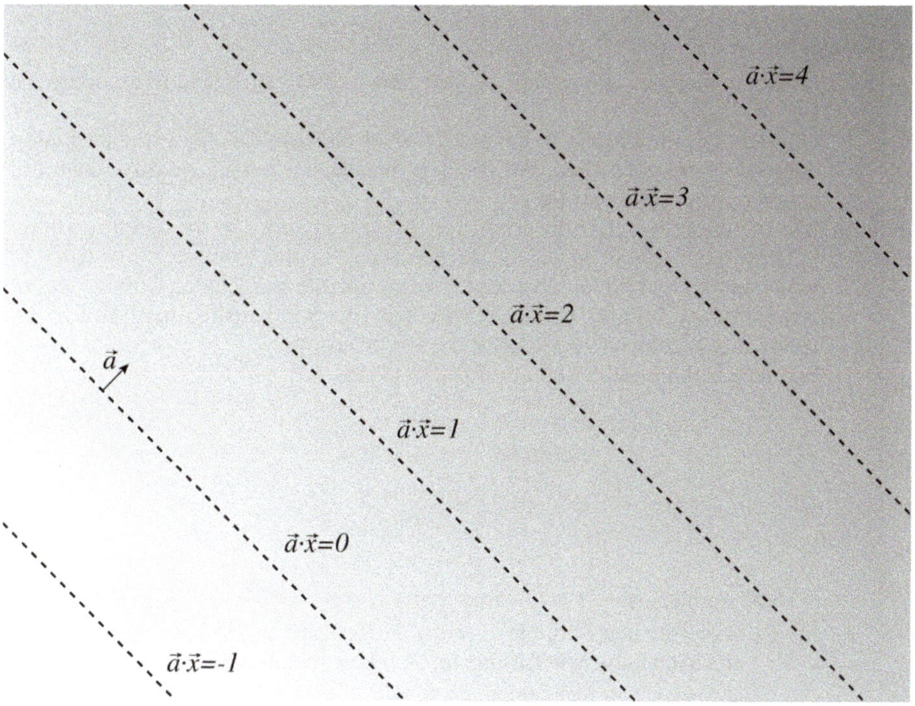

Figure 1.3. *Objective function.* An illustrative example in 2-D of the objective function $f(\vec{x}) = \vec{a} \cdot \vec{x}$, with the direction of $\vec{a}$ depicted above. The dashed lines, perpendicular to $\vec{a}$, are surfaces on which the function is constant.

specified by a vector $\vec{a}$ through $f(\vec{x}) = \vec{a} \cdot \vec{x}$. Geometrically, it is useful to view this as a *direction*, specified by $\vec{a}$. Hyperplanes perpendicular to $\vec{a}$ are the surfaces on which the function is *constant*, as depicted in figure 1.3.

The maximum of the function $f(\vec{x})$ thus occurs at the point (or points) in the feasible set $\mathcal{F}$ which are furthest in the direction of $\vec{a}$. We can therefore see, rather intuitively from a geometric perspective, that if the feasible set is bounded in extent, then the maximum will be finite; if on the other hand the feasible set is unbounded, the maximum will depend on whether the direction of $\vec{a}$ coincides with the direction in which $\mathcal{F}$ is unbounded or not. This is illustrated in figure 1.4.

Geometrically, we also understand why LPs can have infinitely many optimal solutions $\vec{x}^*$. This happens, in particular, when the relevant *face* of the feasible set happens to align with the objective function, i.e. such that the normal vector to the face is in the same direction as the vector $\vec{a}$ specifying the objective function. In this case, all vectors $\vec{x}$ on the relevant face of $\mathcal{F}$ will be optimal, and achieve the same optimal value $\alpha = \vec{a} \cdot \vec{x}^*$.

Exercises

1.12 Draw a collection of inequality constraints in 2-D which lead to an infeasible LP.

1.13 Construct geometrically an LP in 2-D which is feasible as far as the inequality constraints are concerned, but which is infeasible when an equality constraint is taken into consideration.

1.14 It is often the case that the constraint $\vec{x} \geqslant \vec{0}$ is included in the definition of an LP, although this is by no means necessary. In this exercise, we will try to understand why such a constraint can in principle be included without loss of generality, by suitably changing variables. We will restrict attention to 2-D, where it is most easy to visualise the key concepts.

(a) Sketch the feasible region of the following LP

$$\begin{aligned} \text{maximise} \quad & 2x_1 + x_2 \\ \text{subject to} \quad & x_1 + x_2 \leqslant 1 \\ & x_1 - x_2 \leqslant 1 \\ & x_1 + x_2 \geqslant -1 \\ & x_1 - x_2 \geqslant -1 \end{aligned}$$

and confirm that it is not true that $\vec{x} \geqslant \vec{0}$ for all $\vec{x} \in \mathcal{F}$.

(b) Consider the new variable $\vec{x}' = \vec{x} + \vec{1}$. Rewrite the LP from (a) in terms of $\vec{x}'$, and sketch the new feasible set $\mathcal{F}'$. Show therefore that $\vec{x}' \geqslant \vec{0}$ within $\mathcal{F}'$.

*This shows that sometimesa simple **shift** of the variables of the problem make them positive without loss of generality.*

(c) Sketch the feasible region of the following LP

$$\begin{aligned} \text{maximise} \quad & -3x_1 + x_2 \\ \text{subject to} \quad & x_1 + x_2 \geqslant 0 \\ & x_1 - x_2 \geqslant 0 \end{aligned}$$

and confirm that it is not true that $\vec{x} \geqslant \vec{0}$ for all $\vec{x} \in \mathcal{F}$.

(d) Consider the new variable $\vec{x}'$ such that

$$x_1' = \frac{x_1 - x_2}{\sqrt{2}},$$
$$x_2' = \frac{x_1 + x_2}{\sqrt{2}}.$$

Rewrite the LP from (c) in terms of $\vec{x}'$, and sketch the new feasible set $\mathcal{F}'$. Show therefore that $\vec{x}' \geqslant \vec{0}$ within $\mathcal{F}'$. Furthermore, show that this change of variables is an *orthogonal* transformation (that is, it is a real matrix that preserves the inner product between vectors) with determinant 1.

*This shows that sometimes a simple **rotation** of the variables of the problem make them positive without loss of generality.*

(e) Sketch the feasible region of the following LP

$$\begin{aligned} \text{maximise} \quad & x_1 + x_2 \\ \text{subject to} \quad & 2x_1 + x_2 \leqslant 2 \\ & x_1 + 2x_2 \leqslant 2 \end{aligned}$$

and confirm that it is not true that $\vec{x} \geqslant \vec{0}$ for all $\vec{x} \in \mathcal{F}$. Use the sketch to explain why no simple shift, rotation or reflection of the variables will be able to produce new variables $\vec{x}'$ such that $\vec{x}' \geqslant \vec{0}$.

(f) Consider the new variable $\vec{x}'$ such that

$$x_1' = \frac{-2x_1 + x_2 + 2}{3},$$
$$x_2' = \frac{x_1 - 2x_2 + 2}{3}.$$

Rewrite the LP from (e) in terms of $\vec{x}'$, and sketch the new feasible set $\mathcal{F}'$. Show therefore that $\vec{x}' \geqslant \vec{0}$ within $\mathcal{F}'$.

(g) Write the transformation from (f) in the form $\vec{x}' = S\vec{x} + \vec{t}$. Confirm that S is not an orthogonal transformation, but is invertible.

This shows that, in general, it is always possible to find an invertible ***affine*** *transformation – i.e. a shift and an invertible linear transformation – that transforms the feasible set so that it is contained inside the positive quadrant. Can you think of any exceptions to this?*

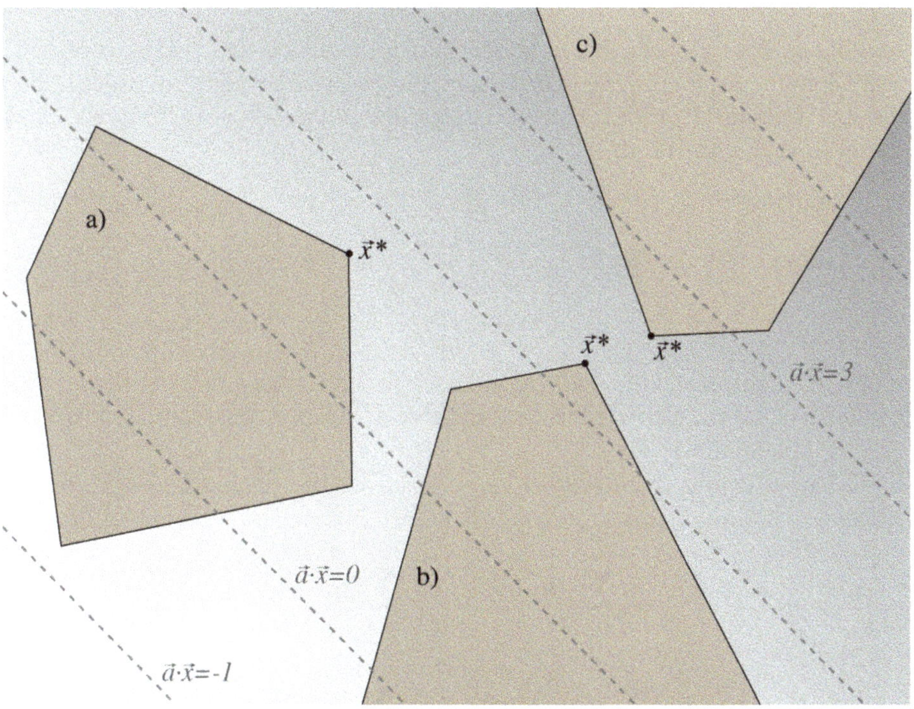

Figure 1.4. *Graphical solutions to LPs.* An illustrative example in 2-D of graphically solving an LP. We depict the feasible sets of different LPs which share the same objective function. In (a) the feasible set is a polytope, with optimal solution $\vec{x}^*$. In (b) the feasible set is unbounded, however, maximising the objective function still leads to a finite solution, attained at $\vec{x}^*$. In (c) the feasible set is again unbounded, and in this case has no maximum value, since it increases indefinitely in the direction of the objective function. If however, we were to consider a minimisation problem over this set (with the same objective function), then the minimum is finite and attained at $\vec{x}^*$.

1.3 Duality

We now introduce an extremely important aspect of the theory of linear programs—*duality*. Every linear program has an alternative formulation, which is known as the *dual problem*. The importance of the dual problem, as will be seen, is that any feasible point it defines provides an upper bound to the original linear program. In order to differentiate the original problem from the dual problem, from now on the original problem will be referred to as the *primal* LP.

As we noted above, it is rather straightforward to find lower bounds on the optimal value of a linear program, since any feasible point $\vec{x} \in \mathcal{F}$ provides such a bound. What is less obvious at first sight, is how to find upper bounds, and this is the first key motivation for introducing the dual problem.

The dual formulation is in reality much more useful than just this. Under some mild assumptions, that will be presented later on, the optimal value of the dual LP is guaranteed to be equal to the optimal value of the primal LP. It is for this reason that we refer to this as the dual formulation, as it provides a completely novel way of expressing the optimisation problem of interest, with wide ranging implications, both from a calculational perspective, as well as from a conceptual perspective.

In what follows we will derive the general form of the dual LP associated to a primal LP in the form (1.3), i.e. when using the vectorial notation. In order to do so, let us first rewrite (1.3) and associate *Lagrange multipliers*—also known as *dual variables* to each constraint:

$$\text{maximise} \quad \vec{a} \cdot \vec{x} \tag{1.26a}$$

$$\text{subject to} \quad \vec{r}_i \cdot \vec{x} = b_i \qquad i = 1, \dots, m \qquad : y_i \tag{1.26b}$$

$$\vec{s}_j \cdot \vec{x} \leqslant c_j \qquad j = 1, \dots, n \qquad : z_j \tag{1.26c}$$

where we have displayed the dual variables (in red) after the colon on the right-hand side of each equation, i.e. y_i is the dual variable associated to the equality constraint $\vec{r}_i \cdot \vec{x} = b_i$, and similarly for z_j.

Let us now define the so-called *Lagrangian* of the problem as the following function

$$\mathcal{L} = \vec{a} \cdot \vec{x} + \sum_{i=1}^{m} y_i(b_i - \vec{r}_i \cdot \vec{x}) + \sum_{j=1}^{n} z_j(c_j - \vec{s}_j \cdot \vec{x}) \tag{1.27a}$$

$$= \left(\vec{a} - \sum_{i=1}^{m} y_i\vec{r}_i - \sum_{j=1}^{n} z_j\vec{s}_j\right) \cdot \vec{x} + \sum_{i=1}^{m} y_i b_i + \sum_{j=1}^{n} z_j c_j, \tag{1.27b}$$

The Lagrangian is constructed by adding additional terms to the objective function, where the additional terms are formed by multiplying the constraints with the associated dual variables. The logic of this construction is that these are *the most general linear functions* of the constraints that can be added—equivalent to an arbitrary scaling. The utility of this will become apparent in the following.

Although not written explicitly, the Lagrangian should be considered as a function of the primal variable $\vec{x}$, and all of the dual variables y_i and z_j.

A priori the dual variables that appear in the Lagrangian are arbitrary. However, it will prove useful to impose some carefully chosen constraints on these variables. In particular, we will first impose that

$$z_j \geqslant 0 \quad \forall j. \tag{1.28}$$

In order to understand why these constraints are imposed, let us consider the value of the Lagrangian when $\vec{x}$ is taken to be a feasible point of the primal LP, $\vec{x} \in \mathcal{F}$. Focusing on (1.27), independent of whether we impose the above constraint or not, notice first that the second term vanishes, since $b_i - \vec{r}_i \cdot \vec{x} = 0$ for all i, from (1.26b). Now, upon imposing the constraint (1.28), the third term will necessarily be *nonnegative*, which follows from (1.26c).

The above implies, in particular, that we have the important property

$$\mathcal{L} \geqslant \vec{a} \cdot \vec{x} \quad \text{for all } \vec{x} \in \mathcal{F} \text{ whenever } z_i \geqslant 0 \ \forall i. \tag{1.29}$$

That is, the Lagrangian is seen to upper bound the value of the objective function for all feasible points $\vec{x}$ when the restriction (1.28) is imposed on the dual variables.

We can go one step further, by focusing on (1.27b). Notice that it is possible to make the Lagrangian *independent of the primal variable* $\vec{x}$ if we further restrict our attention to dual variables which satisfy

$$\vec{a} - \sum_{i=1}^{m} y_i \vec{r}_i - \sum_{j=1}^{n} z_j \vec{s}_j = \vec{0}. \tag{1.30}$$

In this case, the Lagrangian simply equals

$$\mathcal{L} = \sum_{i=1}^{m} y_i b_i + \sum_{j=1}^{n} z_j c_j. \tag{1.31}$$

Because of the previous constraint/restriction on the dual variables, the Lagrangian remains larger than the objective function when evaluated at any feasible primal point, even though it is now independent of the primal variable in this subspace of dual variables. That is, the Lagrangian is constant as far as the primal variables are concerned when the set of dual variables is restricted to satisfy (1.30). Moreover, because the dual variables are simultaneously restricted to satisfy (1.28), we are guaranteed that the value of the Lagrangian is always larger than the value of the primal objective function.

The final key observation is then to realise that the *tightest upper bound on the primal objective function* is obtained by minimising the Lagrangian over the dual variables subject to the carefully chosen constraints (1.28) and (1.30). That is, we arrive at the following optimisation problem, known as the *dual LP*:

$$\text{minimise} \quad \sum_{i=1}^{m} y_i b_i + \sum_{j=1}^{n} z_j c_j \tag{1.32a}$$

$$\text{subject to} \quad \vec{a} - \sum_{i=1}^{m} y_i \vec{r}_i - \sum_{j=1}^{n} z_j \vec{s}_j = \vec{0}, \tag{1.32b}$$

$$z_j \geqslant 0 \quad j = 1, \dots n. \tag{1.32c}$$

We can simplify the form of the dual LP by realising that y_i and z_j can naturally be taken to be components of vectors $\vec{y}$ and $\vec{z}$, living in $\mathbb{R}^m$ and $\mathbb{R}^n$ respectively. The dual objective function is then simply $\vec{b} \cdot \vec{y} + \vec{c} \cdot \vec{z}$. For the constraint, recall that in (1.10) we previously defined B to be the matrix whose rows were $\vec{r}_i^{\mathsf{T}}$. From this it follows that $\sum_i y_i \vec{r}_i = B^{\mathsf{T}} \vec{y}$. Similarly it can be seen that $\sum_j z_j \vec{s}_j = C^{\mathsf{T}} \vec{z}$, where C is defined in (1.11), and is the matrix whose rows are $\vec{s}_j^{\mathsf{T}}$, so that the columns of C^{T} are the vectors $\vec{s}_j$.

Putting everything together, we arrive at the equivalent, but simpler form of the dual LP:

$$\text{minimise} \quad \vec{b} \cdot \vec{y} + \vec{c} \cdot \vec{z} \tag{1.33a}$$

$$\text{subject to} \quad B^{\mathsf{T}} \vec{y} + C^{\mathsf{T}} \vec{z} = \vec{a}, \tag{1.33b}$$

$$\vec{z} \geqslant 0. \tag{1.33c}$$

The structure of the dual is very similar to the primal LP in matrix form as given in (1.12). First of all, it is a constrained optimisation problem, but now a minimisation instead of a maximisation. Second, the objective function, which we will refer to as the *dual objective function*, is linear in the dual variables $\vec{y}$ and $\vec{z}$. Finally, it contains linear equality constraints, and the simplest inequality constraint, $\vec{z} \geqslant 0$. That is, the dual is itself a linear program. The difference is that whereas the primal problem had a single variable $\vec{x}$, in the dual we have ended up with a pair of variables $\vec{y}$ and $\vec{z}$. This apparent difference is in fact purely cosmetic, and it is possible to make the two problems look identical in form. In particular, the dual vectors can be concatenated into a single vector $\vec{y}' = \vec{y} \oplus \vec{z} \in \mathbb{R}^{m+n}$, i.e.

$$\vec{y'} = \begin{bmatrix} \vec{y} \\ \vec{z} \end{bmatrix}. \tag{1.34}$$

Carrying out the rest of the details explicitly is left as an exercise below. In what follows we will continue to instead phrase the dual LP in terms of the pair of variables, and take it as understood that the key structure of a linear program is to have a linear objective function and linear equality and inequality constraints in *any number of variables*[3].

[3] It is useful to recall that $\vec{x}$ was itself just a convenient way of grouping together the k variables x_i. What we see is that when deriving the dual LP, we naturally end up grouping the dual variables into two vectors $\vec{y}$ and $\vec{z}$. Nothing however stops us from further grouping these two into a single vector of dual variables.

One interesting point worth highlighting is how the *dimensions* of the primal and dual problems differ. The primal problem optimises over k variables and involves m equality constraints and n inequality constraints. In contrast, the dual problem optimises over $n + m$ variables and involves k equality constraints and n inequality constraints. Since all of k, n and m can vary independently, this means that there can be big differences between the dimensions of the primal and dual problems, when considered geometrically.

Since the dual problem is itself an LP, this means that we have all of the same ingredients as for the primal LP. When we want to explicitly differentiate between the two, we will prefix with either *primal* or *dual*. When this distinction isn't important, we will drop this prefix.

In particular, any pair of dual variables $(\vec{y}, \vec{z})$ that satisfy the constraints of the dual LP will be said to be *dual feasible*, and the set of all dual feasible variables will be denoted by $\tilde{\mathcal{F}}$. The *dual optimal value* is then compactly

$$\beta = \min_{(\vec{y},\vec{z})\in\tilde{\mathcal{F}}} \vec{b} \cdot \vec{y} + \vec{c} \cdot \vec{z} \tag{1.35}$$

and any pair of variables $(\vec{y}^*, \vec{z}^*)$ that achieve the minimum are called *dual optimal*. As with primal LPs, we will sometimes refer to β simply as '*the value*' of the dual, by which we always mean the optimal value of the dual objective function.

In order to illustrate further how to find the dual of a problem, we will now return to our first example, and obtain its dual LP:

Example 1.6 Dual of marginal problem LP.

In this example, we return to the marginal problem, which was reformulated as an LP in example 1.1. The Lagrangian associated to this problem is

$$\begin{aligned}\mathcal{L} = f(P_{X,Y,Z}(x, y, z)) &+ \sum_{x,y}\nu(x, y)\left(P_{X,Y}(x, y) - \sum_z P_{X,Y,Z}(x, y, z)\right)\\ &+ \sum_{x,z}\nu(x, z)\left(P_{X,Z}(x, z) - \sum_y P_{X,Y,Z}(x, y, z)\right)\\ &+ \sum_{y,z}\nu(y, z)\left(P_{Y,Z}(y, z) - \sum_x P_{X,Y,Z}(x, y, z)\right)\\ &+ \sum_{x,y,z}\omega(x, y, z)P_{X,Y,Z}(x, y, z),\end{aligned} \tag{1.36}$$

where we have introduced sets of dual variables $\{\nu(x, y)\}_{x,y}$, $\{\nu(x, z)\}_{x,z}$, $\{\nu(y, z)\}_{y,z}$ and $\{\omega(x, y, z)\}_{x,y,z}$, associated to the four sets of constraints from the primal problem. For a feasible joint probability distribution $P_{X,Y,Z}(x, y, z)$, the first three of these additional terms will all vanish, and the final term can be made nonnegative by restricting attention to dual variables such that

$$\omega(x, y, z) \geqslant 0 \quad \forall x, y, z. \tag{1.37}$$

Under this restriction, we will have that $\mathcal{L} \geqslant f(P_{X,Y,Z}(x, y, z))$. By further re-arranging, and writing $f(P_{X,Y,Z}(x, y, z))$ explicitly as $f(P_{X,Y,Z}(x, y, z)) = \sum_{x,y,z} a_{x,y,z} P_{X,Y,Z}(x, y, z)$, we arrive at

$$\begin{aligned}\mathcal{L} = & \sum_{x,y,z} (a_{x,y,z} - \nu(x, y) - \nu(x, z) - \nu(y, z) + \omega(x, y, z)) P_{X,Y,Z}(x, y, z) \\ & + \sum_{x,y} \nu(x, y) P_{X,Y}(x, y) + \sum_{x,z} \nu(x, z) P_{X,Z}(x, z) + \sum_{y,z} \nu(y, z) P_{Y,Z}(y, z).\end{aligned} \tag{1.38}$$

We can therefore make the Lagrangian independent of $P_{X,Y,Z}(x, y, z)$—the primal variables—by restricting our attention further to dual variables that satisfy

$$a_{x,y,z} - \nu(x, y) - \nu(x, z) - \nu(y, z) + \omega(x, y, z) = 0 \quad \forall x, y, z. \tag{1.39}$$

We obtain the dual formulation by then finding the best upper bound on the primal objective function, i.e. by minimising the remaining terms in the Lagrangian subject to the set of identified constraints, namely

$$\text{minimise} \quad \sum_{x,y} \nu(x, y) P_{X,Y}(x, y) + \sum_{x,z} \nu(x, z) P_{X,Z}(x, z) + \sum_{y,z} \nu(y, z) P_{Y,Z}(y, z) \tag{1.40a}$$

$$\text{subject to} \quad \nu(x, y) + \nu(x, z) + \nu(x, y) - \omega(x, y, z) = a_{x,y,z} \quad \forall x, y, z, \tag{1.40b}$$

$$\omega(x, y, z) \geqslant 0 \quad \forall x, y, z. \tag{1.40c}$$

Exercises

1.15 Re-express the dual LP (1.33) in terms of the concatenated variable $\vec{y}\,' = \vec{y} \oplus \vec{z} \in \mathbb{R}^{m+n}$.

1.16 Find the dual LP of the Majorisation LP (in maximisation form), from example 1.4.

1.17 Consider the following *minimisation* LP,

$$\begin{aligned} \text{minimise} \quad & \vec{a} \cdot \vec{x} \\ \text{subject to} \quad & B\vec{x} = \vec{b}, \\ & C\vec{x} \geqslant \vec{c}. \end{aligned}$$

Find the dual LP, which will now be a maximisation LP, by introducing dual variables and a Lagrangian which will provide the tightest *lower bound* on the value of the primal LP.

1.4 Slack variables

The dual of the marginal problem in example 1.6 allows us to introduce a new important concept—that of a *slack variable*. These are variables that often arise when switching from a primal to a dual problem, which *do not appear in the objective function*. As such, it is advantageous to *solve* for them, which simplifies the structure of the dual LP, and usually highlights its form in a more natural way.

In the example 1.6 notice that the dual variables $\omega(x, y, z)$ appear *only* within the constraints and *not* in the objective function. We can therefore re-arrange (1.40) in order to solve for these variables, leading to

$$\omega(x, y, z) = \nu(x, y) + \nu(x, z) + \nu(y, z) - a_{x,y,z} \quad \forall x, y, z. \tag{1.41}$$

However, we also have the inequality constraints (1.40b) $\omega(x, y, z) \geqslant 0$, which must not be forgotten about. This will continue to be imposed if we enforce

$$\nu(x, y) + \nu(x, z) + \nu(y, z) \geqslant a_{x,y,z} \quad \forall x, y, z. \tag{1.42}$$

This equation replaces the *pair* of equations (1.40) and (1.40b), combining them into a *single* set of inequality constraints. What this shows is that the *only* role that $\omega(x, y, z)$ was playing was to turn this set of inequality constraints into a set of equality constraints. This is why we refer to them as 'slack variables', as they *pick up the slack* in a set of inequality constraints. After removing these variables, we arrive at the simpler dual formulation of the marginal problem:

$$\begin{aligned} \text{minimise} \quad & \sum_{x,y} \nu(x, y) P_{X,Y}(x, y) + \sum_{x,z} \nu(x, z) P_{X,Z}(x, z) \\ & + \sum_{y,z} \nu(y, z) P_{Y,Z}(y, z) \\ \text{subject to} \quad & \nu(x, y) + \nu(x, z) + \nu(y, z) \geqslant a_{x,y,z} \quad \forall x, y, z. \end{aligned} \tag{1.43}$$

Exercises

1.18. In the main text we show how slack variables can be removed from an LP, which has the effect of converting equality constraints into inequality constraints. Show that, in the converse direction, whenever a problem has inequality constraint(s), it is always possible to add new slack variable(s), satisfying suitable inequality constraint(s), and convert it/them into equality constraint(s).

1.5 Weak and strong duality

In the above we introduced the dual LP in order to obtain the tightest upper bound on the primal optimal value. In particular, to recap, it was shown that

$$\begin{aligned} \alpha &= \vec{a} \cdot \vec{x}^* \\ &= (B^\mathsf{T}\vec{y}^* + C^\mathsf{T}\vec{z}^*) \cdot \vec{x}^* \\ &= \vec{y}^* \cdot B\vec{x}^* + \vec{z}^* \cdot C\vec{x}^* \\ &\leqslant \vec{y}^* \cdot \vec{b} + \vec{z}^* \cdot \vec{c} \\ &= \beta, \end{aligned} \tag{1.44}$$

where the second line follows from the dual constraint (1.33b), the third line follows directly from the scalar product, the fourth line follows from the primal constraints (1.12b) and (1.12c) and the last line follows from the dual objective function (1.33). This important relation is known as *weak duality*.

Notice that any feasible pair of dual variables $\vec{y}$ and $\vec{z}$, i.e. satisfying the constraints of the dual LP, provide an upper bound on the dual optimal value β, since $\vec{b} \cdot \vec{y} + \vec{c} \cdot \vec{z} \geqslant \beta$. Similarly, any feasible primal variable $\vec{x}$—satisfying the constraints of the primal LP—provides a lower bound on the optimal value α, i.e. $\vec{a} \cdot \vec{x} \leqslant \alpha$. Combining these observations with the statement of weak duality, we thus also see that

$$\vec{a} \cdot \vec{x} \leqslant \alpha \leqslant \beta \leqslant \vec{b} \cdot \vec{y} + \vec{c} \cdot \vec{z} \tag{1.45}$$

for all primal feasible points $\vec{x}$ and dual feasible pairs $\vec{y}$ and $\vec{z}$. This is itself an important result that leads to the following realisation: if one finds a set of primal and dual feasible variables such that

$$\vec{a} \cdot \vec{x} = \vec{b} \cdot \vec{y} + \vec{c} \cdot \vec{z} \tag{1.46}$$

then it follows immediately that *they must be primal and dual optimal* respectively. Furthermore, in this case, something interesting has happened—*the values of the primal and dual LPs in fact coincide*, i.e. saturate the bound of weak duality. If this special condition holds, we say that the LP satisfies *strong duality*.

Conceptually, strong duality is very important, and justifies why we refer to the dual LP as *dual*: it really is an equivalent way of arriving at the value of the LP—and here we can refer to *the* value of the LP since the dual value and the primal value are equal.

Crucially, strong duality is not just a theoretical possibility, but is the generic behaviour of LPs: that is, in practice, it is essentially always the case that an LP will satisfy strong duality, unless the problem is set up in a particularly bad way. In particular, it can be shown that:

> *Whenever the primal LP is feasible and bounded, i.e.* $-\infty < \alpha < \infty$, *then strong duality holds, and the value of the primal and dual LPs coincide,* $\alpha = \beta$.

There are two other possibilities, that the primal LP can be either *infeasible* or *unbounded*. In the former case, the primal feasible set $\mathcal{F}$ is empty. In the latter case, the feasible set must be a polyhedral set, rather than a polytope, and the direction of the objective function must align with the direction in which the feasible set is unbounded. In both cases, strong duality will not in general hold.

As will be seen, and as should be fairly intuitive, in any problem that is relevant in this book—and where we want to use duality—or in practice, the LP will always be feasible and bounded, and hence strong duality will hold, and prove to be a powerful tool. The exception to this rule is when considering feasibility problems, where by construction the problem is to determine whether the LP is feasible or not. However, as seen previously, we can recast feasibility LPs as standard LPs, and therefore even in this case duality can still be used as a powerful tool.

Finally, it was also noted that the dual problem is itself an LP. There is in fact nothing special about which problem we call the primal and which the dual, and therefore from the perspective of strong duality, everything can alternatively be phrased in terms of the dual instead. In this case, as long as the dual LP is feasible and bounded, then strong duality will hold.

We will not provide a proof of the fact that if the primal or dual is feasible and bounded then strong duality holds. Proofs can be found in many standard texts on linear programming. Since this book is introductory and focuses on the *use* of linear and semidefinite programming in quantum information science, we rely merely on using it to ensure that the dual formulation is an equivalent expression for the problems considered here.

1.6 Concluding remarks

In this chapter we have covered most of the fundamental aspects of linear programming. The purpose of this was two-fold. First, linear programming is an interesting and relevant topic by itself, and actually many interesting problems in quantum information can be recast as an LP. Second, and more importantly, the aim of this chapter was to serve as an introduction to the main topic of this book, semidefinite programming. As we will see in the next chapter, semidefinite programming is a rather natural extension of LP, that arises when optimising over variables naturally associated with operators, rather than those naturally associated with vectors. Because of this, many of the features and results discussed in this chapter will naturally generalise to SDP.

We hope that at this point you now have a basic familiarity with LPs. In particular, we would like that you keep in mind the following important take-home messages:

- **Definition.** A LP is an optimisation problem (customarily a maximisation) of a real linear function over real variables, subject to various linear equality and inequality constraints—see (1.12).
- **Duality.** Every LP has an associated dual formulation, which is itself an LP (customarily a minimisation)—see equation (1.33).
- **Weak duality.** Every feasible point of the dual LP provides an upper bound on the optimal value of the primal LP. Similarly, every feasible point of the primal LP provides a lower bound on the optimal value of the dual LP—see (1.44).
- **Strong duality.** Under mild assumptions (feasibility and boundedness) the primal and dual optimal values are equal to each other.
- **Representability.** It is sometimes possible to reformulate an optimisation problem with a non-linear objective function and/or non-linear constraints as an LP—see example 1.2.

1.7 Advanced topics

We now go on to cover a couple of more advanced topics. As a reader, if you are just interested in covering just the basics of linear programming, then this section is not as essential as the topics covered above. As such, it can safely be skipped in the first instance. However, in our view, as one gains more familiarity with LPs and SDPs, it should prove useful to revisit these topics, as they will add to the general toolbox of techniques that can be applied to solve problems, and will aid in gaining further insights using linear programming.

1.7.1 Complementary slackness

An important concept which arises from strong duality is the notion of *complementary slackness*, which provides useful information about primal and dual optimal variables, and their relationship. Moreover, as we will see, the complementary slackness conditions furthermore provide us with a set of necessary and sufficient criteria for optimality, which can be used *in place* of solving an LP directly.

Let us return to the Lagrangian associated to the primal problem (1.26), namely

$$\mathcal{L} = \vec{a} \cdot \vec{x} + \sum_{i=1}^{m} y_i(b_i - \vec{r}_i \cdot \vec{x}) + \sum_{j=1}^{n} z_j(c_j - \vec{s}_j \cdot \vec{x}). \tag{1.47}$$

The dual LP (1.32) was obtained by minimising this Lagrangian subject to a number of carefully chosen constraints. When strong duality holds, it tells us that the Lagrangian evaluated using a set of optimal variables is equal to the value of the primal objective function, which is simply $\vec{a} \cdot \vec{x}^*$. This is however just the first term of the Lagrangian, and therefore, the sum of the remaining terms must vanish. The second term vanishes, since $\vec{x}^*$ is feasible. This leaves just the final term. Each term in the sum is however nonnegative, by the constraints of the primal and dual LPs, (1.3c) and (1.32c) respectively. Therefore, when strong duality hold, it follows that

$$z_j^*(c_j - \vec{s}_j \cdot \vec{x}^*) = 0 \quad \forall j. \tag{1.48}$$

These conditions are known as *complementary slackness* conditions. They tell us that there are important *orthogonality relations* between the primal and dual optimal variables. This can most readily be seen by re-expressing (1.48) in vector notation, which reads

$$\vec{z}^* \cdot (\vec{c} - C\vec{x}^*) = 0, \tag{1.49}$$

where we recall that C is the matrix defined in (1.11) whose rows are the transposed vectors $\vec{s}_j^{\mathsf{T}}$. That is, the optimal dual variable $\vec{z}^*$ is orthogonal to the vector $\vec{c} - C\vec{x}^*$ formed from the optimal primal variable $\vec{x}^*$.

We can gain some intuition about the complementary slackness conditions by introducing the notion of whether a constraint is *active* or not. Imagine that we have found an optimal solution of the dual LP, such that some component or components of the optimal dual variable $\vec{z}^*$ do not vanish, e.g. $z_1^* \neq 0$. In this case, the constraint that $z_1 \geqslant 0$ is not active as it did not constrain the optimal value of the dual LP. If we look at the complementary slackness conditions (1.48), in order to satisfy it, $c_1 - \vec{s}_1 \cdot x^* = 0$. This however shows that the constraint $\vec{s}_1 \cdot x^* \leqslant c_1$ from the primal LP is active, i.e. saturated. We can also apply the same reasoning to the inequality constraints of the primal LP, and arrive at similar conclusions.

What this shows us is that whenever a constraint in the primal (or dual) is *not* active—meaning that the inequality is not saturated by the optimal variables—then the constraint on the associated dual (or primal) variable is *necessarily active*. We can view this as showing that the 'slack' in the problem has to be picked up either by the primal or by the dual.

This can also be understood from a simple example. Consider the probability simplex from example 1.5. If we were to add an additional constraint

$$p(0) + p(1) + p(2) \leqslant 2, \tag{1.50}$$

then clearly this constraint is never active, since $p(0) + p(1) + p(2) = 1 < 2$. Thus, including this additional constraint in any LP which optimises over the probability simplex will have no effect on the optimal value. This new constraint constrains the problem in an unessential (or trivial) way.

Now consider how the Lagrangian—and therefore dual LP—of the problem with the additional constraint relates to the original problem. The only difference between these two dual problems would be that the original problem would have fewer dual variables, since there is one fewer constraint in the problem. There is however a different way to think about this: instead of viewing the problem as having one fewer dual variable, we can instead imagine imposing the additional constraint, and at the same time impose that the associated dual variable is zero. This is exactly what happens in complementary slackness—if an optimal dual variable z_j is found that vanishes, *then we know that the corresponding constraint from the primal was in fact non-essential and non-constraining.*

Finally, complementary slackness is important since if we impose it, along with the constraints of the primal and dual problems, a set of conditions which are both *necessary and sufficient for optimality* are obtained. That is, a set of primal and dual variables are jointly optimal if and only if

$$\text{Primal feasible:} \quad B\vec{x}^* = \vec{b}, \qquad C\vec{x}^* \leqslant \vec{c}, \tag{1.51a}$$

$$\text{Dual feasible:} \quad B^\mathsf{T}\vec{y}^* + C^\mathsf{T}\vec{z}^* = \vec{a}, \qquad \vec{z}^* \geqslant 0 \tag{1.51b}$$

$$\text{Complementary slackness:} \quad \vec{z}^* \cdot (\vec{c} - C\vec{x}^*) = 0. \tag{1.51c}$$

These conditions can prove useful. In particular, they provide a method for finding optimal solutions: if the primal and dual feasible sets have been fully characterised, one can try and directly solve the complementary slackness condition. If this can be achieved, we are guaranteed that these variables are optimal. In exercise 1.21 below an example of this will be seen.

Exercises

1.19 In the main text we analysed the situation where the optimal dual variables z_1^* did not vanish (meaning that the first of the dual inequality constraints is not active). We saw that this implied that the associated inequality constraint on the primal variable must be active (i.e. satisfied with equality). Show, on the contrary, that if one of the inequality constraints is not active for an optimal primal variable, then this implies that the associated optimal dual variable vanishes (and hence the inequality constraint is not active).

1.20 In this exercise we will explicitly calculate what happens when non-essential inequality constraints are removed from a simple linear program. Consider the following LP in two variables

$$\text{maximise} \quad x_1 + 2x_2 \tag{1.52a}$$

$$\text{subject to} \quad -1 \leqslant x_1 \leqslant 2, \tag{1.52b}$$

$$-3 \leqslant x_2 \leqslant 4. \tag{1.52c}$$

(a) Solve this LP by inspection to show that the optimal value is $\alpha = 10$.
(b) Write down the Lagrangian associated to this LP, and use it to show that the dual LP is

$$\text{minimise} \quad z_1 + 2z_2 + 3z_3 + 4z_4 \tag{1.53a}$$

$$\text{subject to} \quad z_2 - z_1 = 1, \tag{1.53b}$$

$$z_4 - z_3 = 2, \tag{1.53c}$$

$$z_1 \geqslant 0, \quad z_2 \geqslant 0, \quad z_3 \geqslant 0, \quad z_4 \geqslant 0. \tag{1.53d}$$

(c) Solve the dual by inspection and verify that complementary slackness is satisfied by the optimal primal and dual variables and identify which constraints of the primal and dual LPs are active, and which are inactive.
(d) Consider now the problem formed by ignoring the inactive constraints from the above LP. Write down the primal LP and its Lagrangian. Show that this leads to a trivial dual LP.

1.7.2 Dual form of ℓ_1 and ℓ_∞ norms

In this section we will return to the ℓ_1 and ℓ_∞ norms from example 1.2 and study their dual forms. Using these, in combination with a characterisation of norms in terms of their *dual norms*, we will arrive at an interesting way of characterising the *norm balls* in terms of feasibility LPs.

For the ℓ_1 norm, the Lagrangian associated to the problem (1.6) is

$$\mathcal{L} = \vec{1} \cdot \vec{x} - \vec{u} \cdot (\vec{y} + \vec{x}) - \vec{v} \cdot (\vec{x} - \vec{y}), \tag{1.54a}$$

$$= (\vec{1} - \vec{u} - \vec{v}) \cdot \vec{x} + (\vec{v} - \vec{u}) \cdot \vec{y}, \tag{1.54b}$$

where $\vec{u}$ and $\vec{v}$ are the dual variables associated to the left-hand and right-hand inequalities in (1.6b). Since (1.6) is a minimisation problem, we want $\mathcal{L}$ to be *smaller* than the primal objective function for all feasible primal variables $\vec{x}$, hence we take $\vec{u} \geqslant 0$ and $\vec{v} \geqslant 0$ and note that extra minus signs have been introduced into the Lagrangian compared to how it has been defined previously, in order to make the dual vectors nonnegative. The Lagrangian is made independent of the primal variables by further choosing $\vec{u} + \vec{v} = \vec{1}$. We obtain the dual LP by finding the best lower bound, i.e. by *maximising*, leading to

$$\|\vec{y}\|_1 = \text{maximise} \quad (\vec{v} - \vec{u}) \cdot \vec{y} \tag{1.55a}$$

$$\text{subject to} \quad \vec{u} + \vec{v} = \vec{1}, \tag{1.55b}$$

$$\vec{u} \geqslant 0, \quad \vec{v} \geqslant 0. \tag{1.55c}$$

This is indeed equal to $\|\vec{y}\|_1$ as we can see that *strong duality* holds for vectors $\vec{y}$ of interest. In particular, if all of the entries of $\vec{y}$ are finite, then $\|\vec{y}\|_1 < \infty$, and the primal is clearly feasible, hence strong duality holds. We can learn something more by combining the insight from the dual formulation of the ℓ_1 norm with the expression for the ℓ_1 norm given in exercise 1.6, namely

$$\|\vec{y}\|_1 = \text{maximise} \quad \vec{t} \cdot \vec{y} \tag{1.56a}$$

$$\text{subject to} \quad \|\vec{t}\|_\infty \leqslant 1. \tag{1.56b}$$

By carefully comparing (1.55) and (1.56), we find an interesting method to *characterise the* ℓ_∞ *unit ball*—the set of vectors which have infinity norm less than or equal to unity. The standard definitions for this ball are

$$\mathcal{B}_\infty = \{\vec{t} \mid \|\vec{t}\|_\infty \leqslant 1\}, \tag{1.57a}$$

$$= \left\{\vec{t} \mid -\vec{1} \leqslant \vec{t} \leqslant \vec{1}\right\}, \tag{1.57b}$$

where the second line just states that all components of any vector in the ball must be between −1 and 1.

To arrive at the new characterisation, compare (1.55) and (1.56). Notice that $\vec{v} - \vec{u}$ is playing the role of $\vec{t}$, and hence the constraints on $\vec{u}$ and $\vec{v}$ must ensure that $\|\vec{v} - \vec{u}\|_\infty \leqslant 1$. That is, we are lead to a third—and less obvious—characterisation of the unit ball, namely

$$\mathcal{B}_\infty = \left\{\vec{t} \mid \vec{t} = \vec{v} - \vec{u}, \, \vec{u} + \vec{v} = \vec{1}, \, \vec{u} \geqslant 0, \, \vec{v} \geqslant 0\right\}. \tag{1.58}$$

This form, and the form (1.57b), provide us with two interesting methods to *optimise* over $\mathcal{B}_\infty$, since they show that the ball is *representable by a set of linear inequality and equality constraints*, hence it can be *used inside an LP*, turning a seemingly non-linear constraint on a variable into a set of linear ones.

As an exercise below you will similarly show that the dual form for the ℓ_∞ norm is given by

$$\|\vec{y}\|_\infty = \text{maximise} \quad (\vec{v} - \vec{u}) \cdot \vec{y} \tag{1.59a}$$

$$\text{subject to} \quad (\vec{u} + \vec{v}) \cdot \vec{1} = 1, \tag{1.59b}$$

$$\vec{u} \geqslant 0, \quad \vec{v} \geqslant 0 \tag{1.59c}$$

from which it follows that the *unit ℓ_1 ball*, $\mathcal{B}_1 = \{\vec{t} \mid \|\vec{t}\|_1 \leqslant 1\}$, can be alternatively represented as

$$\mathcal{B}_1 = \left\{\vec{t} \mid \vec{t} = \vec{v} - \vec{u}, (\vec{u} + \vec{v}) \cdot \vec{1} = 1, \vec{u} \geqslant 0, \vec{v} \geqslant 0\right\}. \tag{1.60}$$

This case is more interesting than the case of $\mathcal{B}_\infty$ in some ways, since here it is only through duality that we arrive at a characterisation for $\mathcal{B}_1$ that is representable by a LP. This provides, as above, a method to impose the seemingly non-linear constraint $\|\vec{x}\|_1 \leqslant 1$ inside an LP.

Exercises

1.21 In this exercise we will find the complementary slackness conditions for the ℓ_1 LP in (1.6), and use them to certify optimal variables.

(a) Starting from the Lagrangian (1.54), show that the complementary slackness conditions are

$$u_i^*(y_i - x_i^*) = 0 \quad \forall i, \qquad v_i^*(y_i + x_i^*) = 0 \quad \forall i. \tag{1.61}$$

(b) Show that the following primal and dual variables are feasible for their respective problems:

$$x_i = |y_i|, \qquad u_i = \begin{cases} 1 & \text{if } y_i < 0, \\ 0 & \text{otherwise}, \end{cases} \qquad v_i = \begin{cases} 1 & \text{if } y_i \geqslant 0, \\ 0 & \text{otherwise}. \end{cases} \tag{1.62}$$

(c) Use complementary slackness to show moreover that the variables from (b) are optimal primal and dual variables, and confirm that the primal and dual objective values coincide with each other, and are equal to $\|\vec{y}\|_1$. *We can interpret $\vec{x}^*$ as being the absolute-value-vector of $\vec{y}$, and $\vec{v}^*$ and $\vec{u}^*$ as being projectors onto the positive and negative parts of $\vec{y}$, respectively.*

1.22 In this exercise we will show that the ℓ_1 unit ball is given by (1.60).

(a) Write down the Lagrangian associated to the primal LP for the ℓ_∞ norm given in (1.7), and use it to derive the dual LP as given in (1.59).

(b) Using the fact that

$$\begin{aligned} \|\vec{y}\|_\infty = \text{maximise} \quad & \vec{t} \cdot \vec{y} \\ \text{subject to} \quad & \|\vec{t}\|_1 \leqslant 1, \end{aligned}$$

show that the ℓ_1 unit ball is given by (1.60).

1.23 Derive the dual of the LP for the ℓ_1 norm found in exercise 1.6. How does this relate to the 3 other LPs, (1.6), (1.8) and (1.55), found for the ℓ_1 norm so far?

1.24 We can use the LP representations of $\mathcal{B}_1$ and $\mathcal{B}_\infty$ to arrive at alternative relaxations of the majorisation feasibility LP from example 1.3. In particular, we can introduce a new variable $\vec{q}'$, which is an approximation to $\vec{q}$, and minimise the *distance* between $\vec{q}$ and $\vec{q}'$, where the distance can be either $\|\vec{q} - \vec{q}'\|_1$ or $\|\vec{q} - \vec{q}'\|_\infty$. Explicitly, the two relaxations are

$$\begin{aligned} \text{minimise} \quad & t \\ \text{subject to} \quad & D\vec{p} = \vec{q}\,', \\ & \|\vec{q} - \vec{q}\,'\|_1 \leqslant t, \\ & \sum_x D_{xy} = 1 \ \forall \ y, \\ & \sum_y D_{xy} = 1 \ \forall \ x, \\ & D_{xy} \geqslant 0 \ \forall \ x, y. \end{aligned}$$

$$\begin{aligned} \text{minimise} \quad & t \\ \text{subject to} \quad & D\vec{p} = \vec{q}\,', \\ & \|\vec{q} - \vec{q}\,'\|_\infty \leqslant t, \\ & \sum_x D_{xy} = 1 \ \forall \ y, \\ & \sum_y D_{xy} = 1 \ \forall \ x, \\ & D_{xy} \geqslant 0 \ \forall \ x, y. \end{aligned}$$

Use the characterisations of $\mathcal{B}_1$ and $\mathcal{B}_\infty$ to express both of these problems as LPs.

1.8 Further reading

- Boyd S and Vandenberghe L 2004 *Convex Optimization* (Cambridge: Cambridge University Press) https://web.stanford.edu/~boyd/cvxbook/

IOP Publishing

Semidefinite Programming in Quantum Information Science

Paul Skrzypczyk and Daniel Cavalcanti

Chapter 2

Semidefinite programming

In this chapter we will cover the fundamental aspects of semidefinite programming, which will be seen to be closely related to the fundamental aspects of linear programming covered in the previous chapter. Whereas in the context of linear programming the basic objects of interest were *real vectors*, in the context of semidefinite programming, the basic objects of interest are *Hermitian operators*. We will see later in this chapter that linear programming can in fact be viewed as a special instance of semidefinite programming, which also explains why the two are so intimately related to each other.

2.1 Primal semidefinite programs

A semidefinite program (SDP[1]) is a constrained optimisation problem in an *operator variable*, that we will denote X, and colour in orange. This is a *Hermitian* operator, such that $X^\dagger = X$, and acts on a *complex* vector space, which we will take to be finite dimensional throughout this book. The objective function is a *real* linear function in X, which can always be written as $\text{tr}(AX)$, for some Hermitian operator A. Note that since we consider *exclusively* Hermitian operators, the word 'Hermitian' will almost always be omitted throughout this chapter and book[2].

As with linear programs (LPs), in the context of an SDP the operator X is required to satisfy a number of linear equality and inequality constraints, of the form $\Phi_i(X) = B_i$ for $i = 1, \ldots, m$ and $\Gamma_j(X) \preccurlyeq C_j$ for $j = 1\ldots, n$, where all of the B_i and C_j are Hermitian operators of arbitrary finite dimension, and where $\Phi_i(\cdot)$ and $\Gamma_j(\cdot)$ are linear maps that are *hermiticity-preserving*, which means that $\Phi_i(X)$ and $\Gamma_j(X)$ are Hermitian operators whenever X is Hermitian. Throughout this book we will use

[1] We will use the acronym to denote both semidefinite program and semidefinite programming.

[2] This is similar to the fact that we only considered real vectors in chapter 1, and referred to these simply as 'vectors' rather than the more precise 'real vectors'.

 © IOP Publishing Ltd 2023

the symbol $\preccurlyeq$ (similarly $\succcurlyeq$) to denote an *operator inequality*, i.e. $A \succcurlyeq B$ means that the operator $B - A$ must be *positive semidefinite*, i.e. such that all of its eigenvalues are nonnegative.

Putting these ingredients together, an SDP can be written as

$$\text{maximise} \quad \text{tr}(AX) \tag{2.1a}$$

$$\text{subject to} \quad \Phi_i(X) = B_i \qquad i = 1, \dots, m, \tag{2.1b}$$

$$\Gamma_j(X) \preccurlyeq C_j \qquad j = 1\dots, n. \tag{2.1c}$$

Here, as in the previous chapter, we will assume without loss of generality that the (primal) SDP in this general form is a *maximisation* problem, since any minimisation problem can be expressed as a maximisation by appropriately introducing minus signs, as shown in exercise 2.2 below. In practice, we will encounter naturally both minimisation and maximisation problems and will treat the two on equal footing.

Just as with linear programs, the elements that arise in an SDP—the objective function, the constraint functions, the bounds—fall broadly into two classes, those which specify the problem and its general structure, and the *input data*, which specify the particular instance of the problem. As before, it is insightful to think in analogy to how a piece of code would be written; code will often require input data to run; the lines of code then specify the problem, while the input to the code specifies the instance of the problem. As with LPs in the previous chapter, we will see that which parts of an SDP are specified by the problem and which parts are specified by the instance can vary greatly, depending upon the problem being studied. Also as in the previous chapter, problem data will always be denoted in blue.

With the above in place, consider now a simple example of a problem which naturally arises in quantum theory, and can be cast as an SDP:

Example 2.1 Maximum eigenvalue of a Hermitian operator.

A problem that naturally arises is to find the maximum eigenvalue of a Hermitian operator H, where we treat H as the input data of the problem, specifying a particular instance of the general maximum-eigenvalue problem. This problem can naturally be cast as the following SDP:

$$\text{maximise} \quad \text{tr}(H\rho) \tag{2.2a}$$

$$\text{subject to} \quad \text{tr}(\rho) = 1, \tag{2.2b}$$

$$\rho \succcurlyeq 0. \tag{2.2c}$$

This program maximises the expected value of H with respect to a quantum state ρ. The optimal value will be achieved when ρ lies in the subspace of H with maximum eigenvalue. If H is non-degenerate this corresponds to the projector onto the eigenvector with maximum eigenvalue.

The main point about the above example is that many problems can in fact be cast as SDPs, and this is where their power lies in many respects. The above problem can of course be solved without resorting to SDPs. However, in practice, it can and does prove useful to be able to identify when a problem can be rephrased or recast as an SDP, as this provides methods to solve analytically or numerically a problem, either exactly or approximately, or to find relevant bounds and approximations.

Since we will pre-empt the later section on duality, the SDP (2.1) will be referred to as the *primal SDP*. An operator X that satisfies all of the constraints in (2.1b) and (2.1c) is said to be *(primal) feasible*, where we will only say 'primal' when this is necessary, and will more generally just say that such an X is feasible. As for LPs, we will denote the set of all (primal) feasible X by $\mathcal{F}$. This set has the crucial property of being convex, with essentially the same proof as given in section 1.1.1. In particular, if X_1 and X_2 are both feasible operators, then $X' = pX_1 + (1 - p)X_2$ will also be feasible, for all $0 \leqslant p \leqslant 1$.

With this notation for the feasible set, we can also write compactly the SDP (2.1) as

$$\alpha = \max_{X \in \mathcal{F}} \operatorname{tr}(AX), \tag{2.3}$$

with the *optimal (primal) value* being α. As with LPs, we will refer to α as just '*the value*' of the SDP when no confusion arises, by which it is always meant the optimal value of the primal objective function. A star will be used to denote the optimal (primal) variable, X^*, such that $\alpha = \operatorname{tr}(AX^*)$. Optimal variables of SDPs do not need to be unique, and in general there can be infinitely many optimal variables, which collectively form a convex set.

As with LPs, it is relatively easy to find lower bounds on the optimal values of SDPs—any feasible point $X \in \mathcal{F}$ providing a lower bound, $\operatorname{tr}(AX) \leqslant \alpha$. We will see shortly that upper bounds on optimal values can be found by using *duality*, just as previously for LPs.

If the feasible set of an SDP happens to be the empty set, $\mathcal{F} = \varnothing$, then the SDP is said to be *infeasible*, and the optimal value is again taken to be $\alpha = -\infty$, the smallest possible value. This naturally leads to the special class of SDPs called *feasibility SDPs*. We will use the same special notation as in (1.19) for feasibility SDPs,

$$\text{find} \quad X \tag{2.4a}$$

$$\text{subject to} \quad \Phi_i(X) = B_i \qquad i = 1, \ldots, m, \tag{2.4b}$$

$$\Gamma_j(X) \preccurlyeq C_j \qquad j = 1 \ldots, n. \tag{2.4c}$$

That is, feasibility problems of the form (2.4) are indeed SDPs, and therefore can be solved. It can nevertheless be useful to recast such feasibility SDPs as standard SDPs, by introducing new variables into the problem which suitably relax the constraints, and enlarge the feasible set. By maximising (or minimising) over these new variables, it is then possible to determine whether the original SDP is feasible or not. We will encounter numerous examples of this in later chapters.

A second example, now of a feasibility problem, that arises in quantum theory, is the following:

Example 2.2 Hamiltonian feasibility problem.
Consider the problem where we have an unknown Hamiltonian H, and are only told the average (expected) energy of a set of quantum states, $\mathrm{tr}(H\rho_i) = E_i$, for $i = 1, \ldots, n$. The problem is to determine whether there is a Hamiltonian H whose ground-state energy is nonnegative, consistent with this data. This leads to the following feasibility SDP:

$$\text{find } H \tag{2.5a}$$

$$\text{subject to } \mathrm{tr}(H\rho_i) = E_i \qquad i = 1, \ldots, n, \tag{2.5b}$$

$$H \succcurlyeq 0. \tag{2.5c}$$

A straightforward way to convert this into a standard optimisation SDP is to relax the problem, and to allow the Hamiltonian H to have negative energy eigenvalues, but to maximise the smallest eigenvalue. This can be achieved by replacing the constraint $H \succcurlyeq 0$ by the constraint $H \succcurlyeq \lambda\mathbb{1}$, where λ is a new variable. Maximising the value of λ searches for the Hamiltonian, consistent with the data, which has the largest ground-state energy. The SDP is

$$\text{maximise } \lambda \tag{2.6a}$$

$$\text{subject to } \mathrm{tr}(H\rho_i) = E_i \qquad i = 1, \ldots, n, \tag{2.6b}$$

$$H \succcurlyeq \lambda\mathbb{1}. \tag{2.6c}$$

If $\lambda^* < 0$, this shows that the feasibility problem (2.5) is infeasible—there is no Hamiltonian with positive ground-state energy consistent with the data. On the other hand, if $\lambda^* \geqslant 0$, then the problem was feasible, and a solution H^* is obtained.

As with LPs, it is also the case that certain problems which do not appear to be SDPs—because of non-linear objective function or non-linear constraints—can in fact be cast as SDPs, by making the correct simple observations. As an example, we see that certain norms on operators which are very important in quantum information can be expressed as SDPs, similar to how the ℓ_1 and ℓ_∞ norms studied in example 1.2 can be cast as LPs:

Example 2.3 SDPs for trace and operator norms.
Recall that for *Hermitian* operators the trace and operator norms are, respectively

$$\|A\|_1 = \sum_i |\lambda_i|, \qquad \|A\|_\infty = \max_i |\lambda_i|, \tag{2.7}$$

where $\{\lambda_i\}$ are the eigenvalues of A.[3] In a very similar fashion to the ℓ_1 and ℓ_∞ norms for vectors, studied in example 1.2, we find that an SDP representation of the trace norm is

$$\|A\|_1 = \text{minimise} \quad \text{tr}(X) \tag{2.8a}$$

$$\text{subject to} \quad -X \preccurlyeq A \preccurlyeq X, \tag{2.8b}$$

while an SDP representation of the operator norm is

$$\|A\|_\infty = \text{minimise} \quad x \tag{2.9a}$$

$$\text{subject to} \quad -x\mathbb{1} \preccurlyeq A \preccurlyeq x\mathbb{1}, \tag{2.9b}$$

where x is a (real) scalar variable.

It is worth noting that these two SDPs bare some resemblance to the LPs for ℓ_1 and ℓ_∞ in (1.6) and (1.7) respectively, but are not as straightforward to analyse.

We also note that the operator norm is closely related to the maximum-eigenvalue problem studied in example 2.1, with the only difference being that here it is the eigenvalue with largest absolute value that is of interest. In example 2.4 we derive the dual SDP for the maximum-eigenvalue problem, which should be compared to (2.9).

The above two examples show that semidefinite programming can be applied to problems which even at first sight do not appear to be SDPs, due to inherent non-linearities in the problem. In fact, it is unknown—and still an active area of research—to understand when and how problems can be cast as SDPs. In this book we will see numerous examples where this general technique is used, transforming problems in ingenious ways, so that they can be cast as SDPs.

Exercises

2.1. In the main text it is claimed that any real linear function of a variable X can be written in the form $f(X) = \text{tr}(AX)$ for a suitably chosen Hermitian operator A. In this exercise, we will prove this.
(a) Show that when A and X are both Hermitian, than $\text{tr}(AX)$ is real.
(b) Assuming that $f(X) = \text{tr}(AX)$, show that $f(|j\rangle\langle i|) = \langle i|A|j\rangle$.
(c) Consider an operator A whose matrix elements are $\langle i|A|j\rangle = f(|j\rangle\langle i|)$. Show that, if f is linear, then $\text{tr}(AX) = f(X)$.

2.2 Consider the SDP (2.1), except assume that the goal is to minimise rather than to maximise the objective function. Show that the optimal value α' of this minimisation SDP can be calculated, up to a minus sign, by solving a maximisation SDP with a different objective function. *This implies that minimisation and maximisation SDPs are equivalent, since it is trivial to correct for the sign of α'.*

[3] We restrict here to *Hermitian* operators, since it will significantly simplify the SDPs derived.

2.3 Show that the optimal value of the SDP in example 2.1 is indeed the maximum eigenvalue of the operator H.

2.4 Show that the minimum eigenvalue of a Hermitian operator H can similarly be cast as an SDP, similar to example 2.1.

2.5 Show that the problem of finding the sum of the k largest eigenvalues of a Hermitian operator H can be cast as an SDP. *Note here that we treat k as being input data into the problem, specifying how many eigenvalues to sum. Hint: Consider starting from example* 2.1, *and thinking about how the properties of ρ should change*.

2.6 The trace norm and the operator norm are *dual norms*. This means that an alternative expression for the trace norm is (compare to (1.8))

$$\|A\|_1 = \text{maximise} \quad \operatorname{tr}(AX) \tag{2.10a}$$

$$\text{subject to} \quad \|X\|_\infty \leqslant 1. \tag{2.10b}$$

Using the SDP representation for the operator norm in (2.9), show that the above expression for the trace norm—which is not an SDP as written due to the non-linear constraint—can be recast as the following SDP:

$$\|A\|_1 = \text{maximise} \quad \operatorname{tr}(AX) \tag{2.11a}$$

$$\text{subject to} \quad -\mathbb{1} \leqslant X \leqslant \mathbb{1}. \tag{2.11b}$$

2.7 Let us denote by $A^{(+)}$ and $A^{(-)}$ the positive and negative parts of a Hermitian operator A, respectively. That is, writing $A = \sum_i \lambda_i |\lambda_i\rangle\langle\lambda_i|$ in spectral decomposition,

$$A^{(+)} = \sum_{i:\lambda_i \geqslant 0} \lambda_i |\lambda_i\rangle\langle\lambda_i|, \quad A^{(-)} = \sum_{i:\lambda_i < 0} \lambda_i |\lambda_i\rangle\langle\lambda_i|. \tag{2.12}$$

Show that

$$X^* = A^{(+)} - A^{(-)} \tag{2.13}$$

is a feasible variable for the SDP (2.8), and that it achieves the value $\operatorname{tr}(X^*) = \|A\|_1$, which implies it is also an optimal variable.

2.8 Find an optimal operator X^* for the SDP (2.11), constructed from the projectors onto the positive and negative parts of A,

$$\Pi^{(+)} = \sum_{i:\lambda_i \geqslant 0} |\lambda_i\rangle\langle\lambda_i|, \quad \Pi^{(-)} = \sum_{i:\lambda_i < 0} |\lambda_i\rangle\langle\lambda_i|. \tag{2.14}$$

2.2 Duality

In the previous chapter we saw that every LP has a dual formulation. The same is also true more generally in the context of semidefinite programming. Every primal SDP too has a dual SDP formulation, the optimal value of which provides an upper bound on the optimal value of the primal SDP. Moreover, given mild assumptions—which are satisfied by almost all of the problems of interest in quantum information science—the optimal values of the primal and the dual SDPs coincide.

In what follows we will again explicitly show how the dual SDP formulation of (2.1) can be obtained, using the associated Lagrangian of the optimisation problem, and discuss when the two problems are guaranteed to have the same solution, known as *strong duality*. We will mostly reproduce all of the steps of deriving the dual of an LP that were went through in the previous chapter. There are a number of reasons for going into detail again here. First, it is important to practice the general method of obtaining a dual problem from a primal one. Second, we will gain insight by seeing how the procedure for LPs translates across so directly for SDPs.

Let us associate a Lagrange multiplier, also called a *dual variable*, Y_i for $i = 1, \ldots, m$ to each equality constraint (2.1b) and Z_j for $j = 1, \ldots, n$ to each inequality constraint (2.1c). Each of these dual variables will be taken to be Hermitian, such that $Y_i^\dagger = Y_i$ and $Z_j^\dagger = Z_j$. Using the same notation as in the previous chapter, the primal SDP with dual variables expressed after the colon is:

$$\text{maximise} \quad \text{tr}(AX) \tag{2.15a}$$

$$\text{subject to} \quad \Phi_i(X) = B_i, \quad i = 1, \ldots, m \quad : Y_i \tag{2.15b}$$

$$\Gamma_j(X) \preccurlyeq C_j, \quad j = 1 \ldots, n \quad : Z_j. \tag{2.15c}$$

We then define the following Lagrangian function for the SDP,

$$\mathcal{L} = \text{tr}(AX) + \sum_{i=1}^{m} \text{tr}\{Y_i[B_i - \Phi_i(X)]\} + \sum_{j=1}^{n} \text{tr}\{Z_j[C_j - \Gamma_j(X)]\}, \tag{2.16}$$

$$= \text{tr}\{X[A - \sum_{i=1}^{m} \Phi_i^\dagger(Y_i) - \sum_{j=1}^{n} \Gamma_j^\dagger(Z_j)]\} + \sum_{i=1}^{m} \text{tr}(Y_i B_i) + \sum_{j=1}^{n} \text{tr}(Z_i C_i), \tag{2.17}$$

where the maps $\Phi_i^\dagger(\cdot)$ and $\Gamma_j^\dagger(\cdot)$ are the adjoint maps of $\Phi_i(\cdot)$ and $\Gamma_j(\cdot)$ respectively[4]. That is, we obtain the Lagrangian by adding to the objective function two sets of new terms, which are formed by multiplying the constraints with the dual variables, and taking a trace. The logic for this construction, just as for the case of LPs, is that these are the most general *real linear functions of the constraints* that can be added. As previously, the utility of this will become apparent in the following, and we also emphasise that the Lagrangian should be considered as a function of all of the primal and dual variables, even though it is not explicitly written as such.

We will now use the Lagrangian to place carefully chosen constraints on the dual variables, which up until this stage are arbitrary, apart from being Hermitian. First, we will demand that

$$Z_j \succcurlyeq 0 \quad \forall j. \tag{2.18}$$

[4] Recall that the adjoint map $\Lambda^\dagger(\cdot)$ of $\Lambda(\cdot)$ is the unique map defined such that $\text{tr}(\Lambda(X)Y) = \text{tr}(X\Lambda^\dagger(Y)) \;\forall\; X, Y$.

The reason for imposing these constraints is that when they are in place, the Lagrangian is never smaller than the value of the objective function for any feasible variable X:

$$\mathcal{L} \geqslant \operatorname{tr}(AX) \quad \text{for all } X \in \mathcal{F} \text{ whenever } Z_i \succcurlyeq 0 \ \forall i. \tag{2.19}$$

This can be seen to be true by inspection of the Lagrangian (2.16); the first sum of terms vanishes whenever X is feasible, while the second sum will contain only terms which are nonnegative, since for any pair of positive semidefinite operators A and B, we have that $\operatorname{tr}(AB) \geqslant 0$ (see exercise 2.9). The constraints $Z_i \succcurlyeq 0$ precisely ensure that all operators that appear in the sum are products of positive semidefinite operators, and hence the bound follows.

The second constraint we will impose upon the dual variables arises from inspecting (2.17). Notice that the Lagrangian can be made independent of the primal variable X if we further restrict to dual variables that satisfy

$$A - \sum_{i=1}^{m} \Phi_i^\dagger(Y_i) - \sum_{j=1}^{n} \Gamma_j^\dagger(Z_j) = 0, \tag{2.20}$$

i.e. such that the term multiplying X identically vanishes. Thus, the value of the Lagrangian, when evaluated for a feasible primal variable, and for dual variables which satisfy these constraints becomes equal to

$$\mathcal{L} = \sum_{i=1}^{m} \operatorname{tr}(Y_i B_i) + \sum_{j=1}^{n} \operatorname{tr}(Z_i C_i). \tag{2.21}$$

Crucially, because of the first constraint imposed on the dual variables, we know that this value will always be larger than the value of the primal objective function. Finally, we now see that the tightest upper bound on this value can be obtained by minimising the Lagrangian over all dual variables that satisfy all of the above carefully chosen constraints. That is, we arrive at the following *dual optimisation problem*

$$\text{minimise} \quad \sum_{i=1}^{m} \operatorname{tr}(Y_i B_i) + \sum_{j=1}^{n} \operatorname{tr}(Z_i C_i) \tag{2.22a}$$

$$\text{subject to} \quad A - \sum_{i=1}^{m} \Phi_i^\dagger(Y_i) - \sum_{j=1}^{n} \Gamma_j^\dagger(Z_j) = 0, \tag{2.22b}$$

$$Z_j \succcurlyeq 0 \qquad j = 1, \ldots, n. \tag{2.22c}$$

We can notice that, as expected, the dual problem is itself an SDP. Indeed, the objective function is a linear function of the dual variables, which satisfy a number of linear equality and inequality constraints. Just as we saw for LPs, the *number* of dual variables has increased, compared to the single primal variable. This is however once again a purely cosmetic feature, and it is possible to combine all variables into a single operator variable, which we leave as an exercise below.

As with LPs, the *dual feasible set* $\tilde{\mathcal{F}}$ is defined as the set of all dual variables $(Y_1, \dots, Y_m, Z_1, \dots, Z_n)$ that satisfy the constraints (2.22b) and (2.22c). We refer to the dual SDP as being either feasible or infeasible, depending upon whether this set is empty or not. We will also denote the *optimal value* of the dual problem by β, and can succinctly write the dual as

$$\beta = \min_{\substack{(Y_1, \dots, Y_m, Z_1, \dots, Z_n) \\ \in \tilde{\mathcal{F}}}} \sum_{i=1}^{m} \mathrm{tr}(Y_i B_i) + \sum_{j=1}^{n} \mathrm{tr}(Z_i C_i). \tag{2.23}$$

We can apply duality to example 2.1, which leads to a second natural way of expressing the maximum eigenvalue of an operator as an optimisation problem:

Example 2.4 Dual form of maximum eigenvalue of a Hermitian operator.

We will now derive the dual SDP for the maximum eigenvalue of a Hermitian operator, considered in example 2.1. The first constraint, (2.2b), that $\mathrm{tr}(\rho) = 1$, involves a *scalar*, and hence we need a *scalar dual variable*, which will be called y. The requirement that it should be Hermitian reduces here to requiring that y is *real*. We will associate the dual variable Z with the second constraint, (2.2c), requiring that $\rho \succcurlyeq 0$. The Lagrangian of the problem is therefore

$$\begin{aligned} \mathcal{L} &= \mathrm{tr}(H\rho) + y[1 - \mathrm{tr}(\rho)] + \mathrm{tr}(Z\rho) \\ \mathcal{L} &= \mathrm{tr}[(H - y\mathbb{1} + Z)\rho] + y. \end{aligned} \tag{2.24}$$

We see that $Z \succcurlyeq 0$ should be imposed in order for the last term to be nonnegative, and $H - y\mathbb{1} + Z = 0$ in order for the Lagrangian to become independent of the primal variable ρ. In this case, the Lagrangian becomes simply $\mathcal{L} = y$. We thus arrive at the following dual formulation:

$$\begin{aligned} \text{minimise} \quad & y \\ \text{subject to} \quad & y\mathbb{1} - Z = H, \\ & Z \succcurlyeq 0. \end{aligned} \tag{2.25}$$

We can notice further that the objective function does not depend upon the dual variable Z. Just as was seen in section 1.4 for the case of LPs, SDPs can also have *slack variables*, and here Z is playing the role of such a slack variable, converting what should be a single inequality constraint into an equality constraint with an associated inequality constraint. In particular, we can use (2.25) to solve for Z, giving

$$Z = y\mathbb{1} - H. \tag{2.26}$$

Z however needs to be positive semidefinite, hence $y\mathbb{1} - H \succcurlyeq 0$ needs to be imposed. The final, simplified dual SDP is therefore

$$\text{minimise} \quad y \tag{2.28a}$$

$$\text{subject to} \quad y\mathbb{1} \succcurlyeq H. \tag{2.28b}$$

A little thought shows that this is indeed equal to the maximum eigenvalue of H. In particular, we can notice that $\mathbb{1}$ and H can always be diagonalised in the same basis. The constraint is then seen to imply that y has to be larger than or equal to every eigenvalue of H. The minimum y for which this is true is the maximum eigenvalue of H.

Exercises

2.9 Show that if A and B are both positive semidefinite operators, $A \succcurlyeq 0$ and $B \succcurlyeq 0$, then $\mathrm{tr}(AB) \geqslant 0$.

2.10 In the main text it is claimed that all of the dual variables can always be combined into a single dual variable. Show how this can be done, and write down the dual SDP in terms of a single variable. *Hint: You will want to consider how to concatenate variables into a single variable by forming a block-diagonal operator*.

2.11 In this exercise we will derive the dual SDP of a *minimisation* SDP. Consider the following general form of a minimisation SDP

$$\text{minimise} \quad \mathrm{tr}(AX) \tag{2.29a}$$

$$\text{subject to} \quad \Phi_i(X) = B_i \qquad i = 1, \ldots, m, \tag{2.29b}$$

$$\Gamma_j(X) \succcurlyeq C_j \qquad j = 1, \ldots, n. \tag{2.29c}$$

(a) Write down the Lagrangian for this problem.
(b) Write down the conditions that need to be satisfied by the dual variables in order for the Lagrangian to provide a *lower bound* on the value of the primal SDP when evaluated on primal feasible variables.
(c) Write down the condition that needs to be satisfied by the dual variables in order for the Lagrangian to be independent of the primal variables.
(d) Write down the dual SDP, by maximising the Lagrangian subject to the constraints from parts (b) and (c).

2.12 In example 2.2 we saw the relaxation (2.6) of the Hamiltonian Feasibility Problem. Find the dual SDP of (2.6).

2.13 In exercise 2.4 you showed that the minimum eigenvalue of a Hermitian operator H can be cast as an SDP. Find the dual form of this SDP.

2.14 In exercise 2.5 you showed that finding the k largest eigenvalues of a Hermitian operator H can be cast as an SDP. Find the dual of this SDP.

2.3 Weak and strong duality

We will now return to the important properties of *weak and strong duality*. Weak duality, which always holds, tells us that the optimal value of the dual SDP is an upper bound on the optimal value of the primal. Strong duality, which basically holds in all problems of our interest, tells us that these optimal values coincide.

Concerning weak duality, we can immediately see that

$$\alpha = \operatorname{tr}(AX^*) \tag{2.30a}$$

$$= \operatorname{tr}\left(\left[\sum_i \Phi_i^\dagger(Y_i^*) + \sum_j \Gamma_j^\dagger(Z_j^*)\right] X^*\right) \tag{2.30b}$$

$$= \operatorname{tr}\left(\left[\sum_i Y_i^* \Phi_i(X^*) + \sum_j Z_j^* \Gamma_j(X^*)\right]\right) \tag{2.30c}$$

$$\leqslant \sum_i \operatorname{tr}(Y_i^* B_i) + \sum_j \operatorname{tr}(Z_j^* C_j) \tag{2.30d}$$

$$= \beta. \tag{2.30e}$$

Let us clarify this derivation. The first line is the definition of the optimal primal value. In the second line we have used (2.22b) to substitute for A. The third line is obtained by using the definition of the adjoint maps in reverse, which was used to shift the action of the maps from the dual variables onto the primal variable. In the fourth line we have eliminated the primal variable by making use of the constraints (2.1b) and (2.1c) of the primal SDP, from which the upper bound follows. Finally, the last line is just the definition of the optimal dual value.

Under relatively mild assumptions, a much stronger condition holds, known as *strong duality*, in which case the optimal values of the primal and dual SDPs coincide. This can be viewed as the main reason for calling this the dual SDP, as when strong duality holds, it can be seen as a dual formulation of the same optimisation problem, and in many cases this sheds considerable light on how to think about the problem at hand.

Let us assume that the primal SDP satisfies two natural properties:

(i) The feasible set $\mathcal{F}$ is non-empty, i.e. $\mathcal{F} \neq \varnothing$, and furthermore, there exists a *strictly feasible* solution. That is, if we replace all of the inequality constraints by *strict inequality* constraints,

$$\Gamma_j(X) \prec C_j \quad j = 1, \ldots, n, \qquad X \succ 0, \tag{2.31}$$

while maintaining all of the equality constraints $\Phi_i(X) = B_i$ for $i = 1, \ldots, m$, then it is still possible to find a feasible X. Such an X is said to be in the *relative interior* of the feasible set $\mathcal{F}$. This can be viewed as a type of robustness statement; it says that within the feasible set it should be possible to perturb points and not lose feasibility. Finally, recall then even when interested in feasibility problems, we can always relax to a (standard) optimisation problem, and then apply the above assumption to the relaxation.

(ii) Second, we assume that the optimal value is finite, $\alpha < \infty$. This says that the optimisation should not be unbounded, but should be genuinely constrained, and have a non-trivial maximum value.

If the above two conditions are satisfied, then strong duality holds, and the value of the primal and the dual SDPs coincide. As a summary

Whenever the primal SDP is strictly feasible and bounded, then strong duality holds, and the value of the primal and dual SDPs coincide, $\alpha = \beta$.

In the above, there was nothing special about focusing on the primal problem. We could repeat the entire discussion focusing exclusively on the dual, and it would still be true. That is

If the dual SDP is strictly feasible and bounded, then strong duality also holds.

This is useful, as it is sometimes more straightforward to check these conditions for the dual compared to the primal, or vice versa.

Finally, there is a third condition which can be used, and in practice is most useful of all. If both problems are strictly feasible, then it actually implies that they are bounded. Hence

Strong duality holds if both the primal and the dual problems are strictly feasible.

Just like in the case of strong duality for LPs, we will not provide a proof of strong duality of SDPs in this book. This is not the focus of this book—which is instead about *using* semidefinite programming, rather than deriving the properties they have. Proofs of strong duality can be found in almost all standard texts on the subject, including in those contained at the end of this chapter under *Further reading*.

In order to illustrate these concepts, we will return to the SDP formulations of the trace and operator norm from example 2.3. Strong duality will provide us with alternative SDP representations of these two important norms:

Example 2.5 Dual SDPs for trace and operator norms.

In this example we will return to the trace and operator norms from example 2.3 and derive the dual SDP for the trace norm. This will be used to find an expression for the *unit operator norm ball.* We will state the results for the operator norm, with calculations left as an exercise below. It is particularly interesting to compare what follows to what was seen in the case of vectors, in section 1.7.2.

The Lagrangian for (2.8) is

$$\mathcal{L} = \mathrm{tr}(X) - \mathrm{tr}[Z_1(A + X)] - \mathrm{tr}[Z_2(X - A)] \tag{2.32a}$$

$$= \mathrm{tr}[X(\mathbb{1} - Z_1 - Z_2)] + \mathrm{tr}[(Z_2 - Z_1)A] \tag{2.32b}$$

where we have introduced dual variables $Z_1 \succcurlyeq 0$ and $Z_2 \succcurlyeq 0$ associated to the left-hand and right-hand inequalities in (2.8b) respectively. Note that because the primal problem is a minimisation, minus signs have been introduced for the last two terms in $\mathcal{L}$. We want the Lagrangian to *lower bound* the primal objective function for all feasible X, and this is now the case given that $Z_1 \succcurlyeq 0$ and $Z_2 \succcurlyeq 0$. Making the Lagrangian independent of the primal variables and maximising it leads to the dual SDP for the trace norm,

$$\|A\|_1 = \text{maximise} \quad \text{tr}[(Z_2 - Z_1)A] \tag{2.33a}$$

$$\text{subject to} \quad Z_1 + Z_2 = \mathbb{1}, \tag{2.33b}$$

$$Z_1 \succcurlyeq 0, \quad Z_2 \succcurlyeq 0. \tag{2.33c}$$

It is straightforward to see that this dual is *strictly feasible*, e.g. by taking $Z_1 = Z_2 = \mathbb{1}/2$. The primal problem (2.8) is also strictly feasible by inspection, e.g. by taking $X = \lambda\mathbb{1}$, for any $\lambda > \|A\|_1$. Since both problems are strictly feasible, strong duality holds, and (2.33) is indeed a second formulation of the trace norm.

In exercise 2.11 an alternative SDP for the trace norm was derived, given explicitly in (2.11). Comparing this with the dual formulation (2.33) derived here, we obtain a characterisation of the *unit operator norm ball*, $\mathcal{B}_\infty = \{W \mid \|W\|_\infty \leqslant 1\} = \{W \mid -\mathbb{1} \preccurlyeq W \preccurlyeq \mathbb{1}\}$, given by

$$\mathcal{B}_\infty = \{W \mid W = Z_2 - Z_1, Z_1 + Z_2 = \mathbb{1}, Z_1 \succcurlyeq 0, Z_2 \succcurlyeq 0\}. \tag{2.34}$$

In a similar fashion, this can be repeated for the operator norm. We find that the dual SDP is

$$\|A\|_\infty = \text{maximise} \quad \text{tr}[(Z_2 - Z_1)A] \tag{2.35a}$$

$$\text{subject to} \quad \text{tr}(Z_1 + Z_2) = 1, \tag{2.35b}$$

$$Z_1 \succcurlyeq 0, \quad Z_2 \succcurlyeq 0. \tag{2.35c}$$

This dual SDP formulation can be used to obtain a characterisation of the *unit trace norm ball*, $\mathcal{B}_1 = \{W \mid \|W\|_1 \leqslant 1\}$, given by

$$\mathcal{B}_1 = \{W \mid W = Z_2 - Z_1, \text{tr}(Z_1 + Z_2) = 1, Z_1 \succcurlyeq 0, Z_2 \succcurlyeq 0\}. \tag{2.36}$$

This form is particularly interesting, since in comparison to the operator norm ball, there is no obvious characterisation that can be directly arrived at.

The significance of (2.34) and (2.36) is that they show that the *non-linear* constraints $\|X\|_1 \leqslant 1$ and $\|X\|_\infty \leqslant 1$ can be imposed as constraints *inside SDPs*, something which at first sight is not obvious at all.

The above example hopefully demonstrates that it is possible to check strong duality fairly easily, especially for simple SDPs. In all of the problems studied in this book, and in practice in almost all of the problems that tend to be encountered in quantum information science where semidefinite programming can be applied, we find that strong duality holds, and it is usually straightforward (or relatively easy) to check that this is the case.

It is nevertheless possible to find examples of SDPs where strong duality doesn't hold, and you will find these presented in many standard textbooks on SDPs. The crucial point is that these examples are usually constructed explicitly in order just to show that there are exceptions to the rule of strong duality. Thus, although it is important to understand that strong duality is *not a given*, from here on out we will not dwell much on this point, and will simply verify whenever we present the dual SDP of a problem that strong duality holds (or leave it as an exercise).

Exercises 2.3

2.15 Show that strong duality holds for the maximum eigenvalue SDP of a Hermitian operator from examples 2.1 and 2.4.

2.16 (a) Derive the dual SDP for the operator norm given in (2.35).

(b) Using the fact that the operator norm and trace norm are *dual norms*, find an SDP for the operator norm in a form analogous to (2.10) for the trace norm.

(c) Using your answer to part (b), derive the characterisation of the trace norm ball given in (2.36).

2.17 Starting from example 2.5, find characterisations for the ϵ-balls for the operator and trace norm, i.e. the sets $\{W \mid \|W\|_\infty \leqslant \epsilon\}$ and $\{W \mid \|W\|_1 \leqslant \epsilon\}$. Can the constraints $\|X\|_1 \leqslant \epsilon$ and $\|X\|_\infty \leqslant \epsilon$ also be imposed as constraints inside SDPs?

2.18 Show that optimal dual variables for the SDP (2.33) are

$$Z_1^* = \Pi^{(-)}, \qquad Z_2^* = \Pi^{(+)}, \tag{2.37}$$

where $\Pi^{(+)}$ and $\Pi^{(-)}$ are the projectors onto the positive and negative parts of A, as defined in (2.14). That is, show that these operators are feasible, and that they achieve the value $\|A\|_1$. *This is an important lesson, which shows that we can view the dual SDP as a variational method for finding the projectors onto the positive and negative parts of* A.

2.19 Find optimal dual variables for the SDP (2.35), which achieve the value $\|A\|_\infty$.

2.20 In exercise 2.10 an alternative SDP was derived for the trace norm.

(a) Show that the dual formulation of the SDP (2.11) from this exercise is

$$\|A\|_1 = \text{minimise} \quad \operatorname{tr}(Z_1 + Z_2) \tag{2.38a}$$

$$\text{subject to} \quad A = Z_2 - Z_1, \tag{2.38b}$$

$$Z_1 \succcurlyeq 0, \quad Z_2 \succcurlyeq 0. \tag{2.38c}$$

(b) Explain how this relates to the characterisation of the unit trace norm ball (2.36) from example 2.5 and the trace norm ϵ-ball from exercise 2.17.

2.4 Complementary slackness

As with LPs, *complementary slackness* also arises from strong duality in the case of SDPs, and tells us important information about the optimal primal and dual variables.

As we did in the context of linear programming, our starting point is to return to the Lagrangian from (2.16), namely

$$\mathcal{L} = \operatorname{tr}(AX) + \sum_{i=1}^{m} \operatorname{tr}\{Y_i[B_i - \Phi_i(X)]\} + \sum_{j=1}^{n} \operatorname{tr}\{Z_j[C_j - \Gamma_j(X)]\}. \tag{2.39}$$

When strong duality holds, then $\mathcal{L} = \alpha = \beta$ when we substitute in optimal variables X^*, Y_i^* for $i = 1, \dots, m$ and Z_j^* for $j = 1\dots, n$. This follows from how the dual was constructed, which was to use the Lagrangian to find the best upper bound on the primal objective function. Now, since X^* is by assumption feasible, the second term identically vanishes. Every term in the summation in the third term is always nonnegative by construction, whenever we use feasible (primal and dual) variables. Thus, the only way in which the Lagrangian can equal the primal objective value is if every single one of these terms vanish,

$$\mathrm{tr}\{Z_j^*[C_j - \Gamma_j(X^*)]\} = 0 \quad \text{for } j = 1, \dots, n. \tag{2.40}$$

We can however go one step further. Recall that all $Z_j \succcurlyeq 0$, in order to be dual feasible, and that $C_j - \Gamma_j(X) \succcurlyeq 0$ always holds in order for X to be primal feasible. The only way that $\mathrm{tr}(AB) = 0$, when both $A \succcurlyeq 0$ and $B \succcurlyeq 0$ is if $AB = 0$ (and similarly $BA = 0$). Thus, we see that when strong duality holds then the optimal primal and dual variables satisfy

$$Z_j^*[C_j - \Gamma_j(X^*)] = 0 \quad \text{for } j = 1, \dots, n, \tag{2.41}$$

which are known as the *complementary slackness conditions*. These are important and useful relations between the optimal variables, which can be viewed as *orthogonality relations* between the optimal primal and dual variables. In particular, as with LP, complementary slackness provides us information about when constraints are *active*, and moreover, to what extent they are active.

For example, let us focus on the constraint $Z_1 \succcurlyeq 0$. Imagine that an optimal dual variable Z_1^* is found that is not identically zero, but has a number of non-zero eigenvalues. Let us denote the projector onto the subspace of non-zero eigenvalues, the *support* of Z_1^* by Π_1. What this shows is that the constraint $Z_1^* \succcurlyeq 0$ is only active on the complementary subspace $\mathbb{1} - \Pi_1$, i.e. on the *nullspace* of Z_1^*. Now, since $Z_1^*[C_1 - \Gamma_1(X^*)] = [C_1 - \Gamma_1(X^*)]Z_1^* = 0$, we have that $Z_1^*[C_1 - \Gamma_1(X^*)] - [C_1 - \Gamma_1(X^*)]Z_1^* = 0$, i.e. $[C_1 - \Gamma_1(X^*), Z_1^*] = 0$—they commute, and hence can be diagonalised in the same basis. But then it is straightforward to see, due to the positive semidefiniteness of the operators, that $C_1 - \Gamma_1(X^*)$ must *vanish* on the support of Z_1^*, otherwise it is impossible that their product is zero, hence the support of $C_1 - \Gamma_1(X^*)$ is contained in the nullspace $\mathbb{1} - \Pi_1$ of Z_1^*, and conversely, the support of Z_1^* is seen to be in the nullspace of $C_1 - \Gamma_1(X^*)$.

Exercises

2.21 Primal and dual SDP formulations of the trace norm were given in (2.8) and (2.33) respectively. Optimal primal and dual variables were found in exercises 2.7 and 2.18 respectively. Show that these optimal variables satisfy complementary slackness.

2.5 Linear programs as special instances of semidefinite programs

We finish this section by discussing an important fact—that linear programming can be viewed as a special case of semidefinite programming.

First of all we can see that if all of the constraint of an SDP were written explicitly in terms of the matrix elements of the involved operators, we would end up with a set of *polynomial* constraints in these variables, with the degree of the polynomial depending upon the dimension of the operators. This arises because an operator is positive semidefinite if and only if all of its leading principle minors (determinant of an upper-left sub-matrix) are nonnegative, and these determinants are polynomials in the matrix elements of the operator. We consider the example for 2×2 operator in exercise 2.22 below.

Because of this important realisation, SDPs can also be seen as optimisation problems with a linear objective functions satisfying special types of polynomial constraints, which generalise the linear inequality constraints required to be a linear program.

Second, we can also explicitly rewrite any LP in the language of an SDP. The basic observation is that vectors can be viewed as diagonal operators. Let us assume that we want to re-express an LP given in vector form (as in (1.3)), namely,

$$\text{maximise} \quad \vec{a} \cdot \vec{x} \tag{2.42a}$$

$$\text{subject to} \quad \vec{r}_i \cdot \vec{x} = b_i \qquad i = 1, \ldots, m, \tag{2.42b}$$

$$\vec{s}_j \cdot \vec{x} \leqslant c_j \qquad j = 1, \ldots, n, \tag{2.42c}$$

as an SDP. First consider that we are able to optimise over diagonal operators, and so can encode the variable $\vec{x}$ from the LP in an operator X as

$$X = \sum_i x_i |i\rangle\langle i| = \text{diag}(\vec{x}). \tag{2.43}$$

If we similarly define the diagonal operator $A = \sum_i a_i |i\rangle\langle i| = \text{diag}(\vec{a})$, then the objective function of the LP, $\vec{a} \cdot \vec{x}$, can then be expressed as

$$\vec{a} \cdot \vec{x} = \text{tr}(AX), \tag{2.44}$$

as required for the objective function of an SDP. We can apply the same idea for all of the constraints. In particular, it is possible to take

$$R_i = \sum_k (r_i)_k |k\rangle\langle k| = \text{diag}(\vec{r}_i), \quad S_j = \sum_k (s_j)_k |k\rangle\langle k| = \text{diag}(\vec{s}_j), \tag{2.45}$$

$$\Phi_i(X) = \text{tr}(R_i X), \qquad \Gamma_j(X) = \text{tr}(S_j X), \tag{2.46}$$

$$B_i = b_i, \qquad C_j = c_j, \tag{2.47}$$

which then expresses the constraints of an LP in SDP form. We have thus almost recast the LP as an SDP. The one final ingredient is to further enforce the assumption that X is a diagonal operator. This can be ensured by introducing a further equality constraint. In particular, consider the following map, which is the difference between the *dephasing* and *identity* maps, the former of which sets all off-diagonal elements of an operator to zero,

$$\begin{aligned}\Phi^{(\text{diag})}(X) &= \Phi^{(\text{deph})}(X) - \Phi^{(\text{id})}(X) \\ &= \sum_i |i\rangle\langle i|X|i\rangle\langle i| - X.\end{aligned} \tag{2.48}$$

If an operator X is diagonal, then $\Phi^{(\text{deph})}(X) = X$ and so $\Phi^{(\text{diag})}(X) = 0$. On the other hand, if X has any non-zero off-diagonal elements, then $\Phi^{(\text{diag})}(X) \neq 0$. Thus, adding the constraint

$$\Phi^{(\text{diag})}(X) = 0, \tag{2.49}$$

to an SDP, has the effect of restricting the feasible set $\mathcal{F}$ to be contained within the set of diagonal operators. Hence in this way we can recover an arbitrary LP of the form (2.42) in the form of an SDP, namely

$$\text{maximise} \quad \text{tr}(\text{diag}(\vec{a})X) \tag{2.50a}$$

$$\text{subject to} \quad \text{tr}(\text{diag}(\vec{r}_i)X) = b_i \qquad i = 1, \ldots m, \tag{2.50b}$$

$$\text{tr}(\text{diag}(\vec{s}_j)X) = c_j \qquad i = 1, \ldots m, \tag{2.50c}$$

$$\Phi^{(\text{diag})}(X) = 0. \tag{2.50d}$$

As such, we see that any LP can also be thought of as a semidefinite program. Thus, any result stated about the general theory of semidefinite programming immediately holds for linear programming, and this is one of the main reason for showing that LPs is a special instance of SDPs. In practice, it is often much more useful to identify a problem as a linear program if it is one (rather than leaving it as an SDP), as this can speed numerics, or simplify analytics.

Exercises

2.22. Consider an arbitrary Hermitian 2 × 2 operator, with corresponding matrix

$$X = \begin{pmatrix} x_{11} & x_{12} \\ x_{12}^* & x_{22} \end{pmatrix}, \tag{2.51}$$

where x_{11} and x_{22} are real, while x_{12} is complex.

(a) Show that the eigenvalues of this operator are

$$\lambda_{1,2} = \frac{1}{2}\left(x_{11} + x_{22} \pm \sqrt{(x_{11} - x_{22})^2 + 4|x_{12}|^2}\right). \tag{2.52}$$

(b) Assuming that X is normalised so that $\mathrm{tr}(X) = 1$, show that $X \succcurlyeq 0$ if and only if

$$x_{11}(1 - x_{11}) \geqslant |x_{12}|^2. \tag{2.53}$$

This shows that for a 2 × 2 operator to be positive semidefinite, a non-linear inequality constraint must be satisfied by its matrix elements. This demonstrates how SDPs generalise the linear inequality constraints of LPs to non-linear polynomial constraints.

2.6 Concluding remarks

This brings to a close this chapter on the basics of semidefinite programming, and also the first part of this book, covering the introduction and basics. We hope that at this stage you understand the following key points:

- **Definition.** An SDP is an optimisation problem of a linear function of an operator variable, subject to a number of linear operator equality and inequality constraints—see (2.1).
- **Duality.** Every SDP has a dual formulation, which is also an SDP—see (2.22).
- **Weak and strong duality.** Just like for LPs, we have both weak and strong duality for SDPs. The former means feasible points of the primal and dual SDPs always provide bounds on the optimal values of the other problem (see (2.30)). Strong duality says that if either problem strictly feasible and bounded, or both problems are strictly feasible, then the optimal values of the two coincide.
- **Representability.** It is possible to recast optimisation problems with non-linear objective functions or non-linear constraints as SDPs—see example 2.3.
- **Special case:** LPs are special cases of SDPs, which are a much more general class of optimisation problems.

We now move on to the main part of the book, where our focus is on putting into practice all of the abstract concepts presented up until now, in the context of quantum information science. We will focus on seeing how a wide range of problems, from across the field, can be cast as linear or semidefinite programs, how duality is useful both from a calculational perspective and a conceptual perspective, and how the ability to find SDPs allow for powerful relaxations or approximations of many interesting problems.

2.7 Further reading

- Boyd S and Vandenberghe L 2004 *Convex Optimization* (Cambridge: Cambridge University Press) https://web.stanford.edu/~boyd/cvxbook/

- Watrous J 2018 *Theory of Quantum Information* (Lecture Notes) (Cambridge: Cambridge University Press) section 1.2.3 https://cs.uwaterloo.ca/~watrous/TQI/
- Watrous J 2011 *The Theory of Quantum Information* (Cambridge: Cambridge University Press) ch 7 and 8 https://cs.uwaterloo.ca/~watrous/TQI-notes/
- Sikora J and Varvitsiotis A 2015 *Semidefinite Programming & Quantum Information*, lectures 7–9, presented at Perimeter Institute for Theoretical Physics, Waterloo, Canada https://sites.google.com/site/jamiesikora/teaching/semidefinite-programming-quantum-information

Part II

Semidefinite programming in quantum information science

Chapter 3

Quantum states

In this chapter we begin the second—and main—part of this book, where we explore the use of semidefinite programming in the context of quantum information science. As will be seen, semidefinite programming proves to be a versatile and powerful tool, which finds widespread and varied applications.

One of the first reasons why SDPs are so useful in quantum information is because—mathematically—quantum states are positive semidefinite operators, $\rho \succcurlyeq 0$, which are normalised, $\mathrm{tr}(\rho) = 1$. Being a linear operator inequality and an equality constraint respectively, these conditions can be used as constraints inside SDPs, so that many problems involving optimisation over quantum states can be cast as SDPs. In this chapter we will describe some of these problems. We will discuss problems related to *quantum state estimation*, which refers to the question of estimating the state that a source produces given the outcome statistics of a set of measurements. In the case this set of measurements is tomographically complete, this is equivalent to quantum state tomography—characterising the state emerging from a source. Sometimes, however, the information obtained experimentally is not enough to single out a unique quantum state. With the help of SDPs we can nevertheless characterise the set of states compatible with a given set of experimental observations, and estimate additional (unmeasured) properties of quantum states.

Altogether, these topics will allow us to see how problems involving quantum states can be cast as SDPs, and to use various aspects of the theory of SDPs presented in the previous chapters, in order to gain important insights about these problems.

We will end this chapter with a brief section on the *quantum marginal problem*, which can be viewed as a generalisation of quantum state estimation. In this problem, we consider *multipartite* systems, and are given complete information only about some of their *subsystems*. The goal is then to determine the global state, or properties of it. As will be seen, the tools of semidefinite programming can also be used to solve this problem, and leads to a simple way to see when certain sets of marginals are inconsistent with each other.

doi:10.1088/978-0-7503-3343-6ch3

© IOP Publishing Ltd 2023

3.1 Quantum state estimation

Suppose an experimentalist comes to you and says

> *I have a source of spin-1/2 particles (i.e. qubits) in my laboratory, that I have measured along two different spin directions $\hat{x}$ and $\hat{y}$, obtaining the following expectation values: $\langle \sigma_x \rangle = 0.9$ and $\langle \sigma_y \rangle = 0.5$,*

where σ_x, σ_y and σ_z are the three Pauli operators. Is there any way to know if they are telling the truth or not? More precisely, is it possible to determine whether what they claim to have observed is consistent with the predictions of quantum mechanics? The simple situation above can indeed be checked through basic algebraic means, using the properties of the measurements used. But in more complicated and sophisticated situations, involving, e.g. many measurements with no symmetry relations between them, or high-dimensional quantum system, the task becomes more intricate.

However, this problem can in fact be cast as a simple feasibility SDP. Namely, suppose that we are given the measurement results for a set of N observables $M_1, \ldots, M_N$. Let us assume that we are only told the *average* values of each measurement, $\langle M_1 \rangle = m_1, \ldots, \langle M_N \rangle = m_N$. Our goal is to check if these reported values are compatible with some quantum state ρ. This amounts to solving the following feasibility SDP:

$$\text{find} \quad \rho \tag{3.1a}$$

$$\text{subject to} \quad \mathrm{tr}(M_x \rho) = m_x \qquad x = 1, \ldots, N, \tag{3.1b}$$

$$\mathrm{tr}(\rho) = 1, \tag{3.1c}$$

$$\rho \succcurlyeq 0. \tag{3.1d}$$

There are three possibilities that can arise.

1. **Unique solution.**The set of observables $\{M_x\}$ we consider might have the property of being *tomographically complete*. This means that knowing the results for every measurement in the set is sufficient to uniquely determine the state. Mathematically, the set must have a sufficient number of linearly independent measurements in order to determine all of the matrix elements of the density operator ρ. In this case, the feasible set of the SDP is a unique density operator $\mathcal{F} = \rho^*$.
2. **No solution.** The other extreme case is when there is *no solution*, i.e. there is no quantum state ρ which is able to produce the desired list of expectation values (and the experimentalist was lying or mistaken). In this case, the SDP is *infeasible*, and we see that the feasible set must be empty, $\mathcal{F} = \varnothing$. We will study this in detail in section 3.1.5, where it is shown that the dual SDP can be used to provide us with a certificate, which will convince us that there is no quantum state producing the desired set of expectation values.

3. **Multiple solutions**. The situation in between the above two cases is that there are infinitely many states ρ which produce the given set of expectation values. This means that the set of measurements is not tomographically complete, such that they do not uniquely determine all of the matrix elements of the density operator. On the other hand, the expectation values are consistent with each other. In terms of the SDP, here the feasible set $\mathcal{F}$ will be non-empty, and contain more than a single state.

3.1.1 Trace distance estimation

The above shows that quantum state estimation can be cast as an SDP. We can however use semidefinite programming to dig much deeper into the problem of state estimation. As a first example, imagine that not only do we have the set of measurement results, but also a *target state*, that we believe was prepared, and which was measured to produce the observed statistics. If the statistics are incompatible with this target state we might be interested in understanding *how close* the prepared state was to the target state.

In this section we will see that this question can be answered using semidefinite programming. In particular, we will show that it is possible to find the *best-case* trace distance between any state consistent with the measurement results and some target state.

The key ingredient is to use the primal SDP formulation for the trace norm, as given in (2.8), which we will see can be harnessed here for our needs.

Recall that for a pair of quantum states the trace distance is given by

$$T(\rho, \sigma) = \frac{1}{2}\|\rho - \sigma\|_1. \tag{3.2}$$

This is a non-linear function of ρ and σ, and so cannot be used directly as either the objective function or as a constraint inside an SDP. However, as we saw in example 2.3 in the previous chapter, it can nevertheless be possible to recast problems which appear non-linear as SDPs. We will see here that trace distance estimation is indeed such a problem where the non-linearity is not a road block. The first thing to note is that although non-linear, the trace distance is nevertheless a convex function of either state, such that

$$T(q\rho_0 + (1 - q)\rho_1, \sigma) \leqslant qT(\rho_0, \sigma) + (1 - q)T(\rho_1, \sigma), \tag{3.3}$$

and similarly for the second state.

There are two possibilities that are of potential interest—the *best case* and the *worst case*. In the former, we would want to find the *closest* state to our target state consistent with the measurement results. On the other hand, in the latter case, we would seek to find the state which is furthest away from our target state and still consistent with the results.

The fact that the trace distance is convex means that we will not be able to analyse the *worst case*, but only the best case. Why is this? This is in fact an important lesson about convex optimisation—more general than just linear programs (LPs) and SDPs. As the following sketch illustrates, *convex* objective functions are naturally associated with *minimisation* problems and *concave* objective functions are naturally associated with *maximisation* problems: maximising a convex function we are led to the extremes of the function—to the boundary of the domain. This is a fundamentally different type of optimisation problem compared with finding the optima, which naturally occur at *stationary points.*

Since the trace distance is convex, we can consider problems involving its minimisation, which here would correspond to finding the *minimal distance* or *closest state* to the target state. In particular, the problem we can attempt to solve is

$$\text{minimise} \quad \frac{1}{2}\|\rho - \sigma\|_1 \tag{3.4a}$$

$$\text{subject to} \quad \mathrm{tr}(M_x\rho) = m_x \quad x = 1, \ldots, N, \tag{3.4b}$$

$$\mathrm{tr}(\rho) = 1, \tag{3.4c}$$

$$\rho \geqslant 0. \tag{3.4d}$$

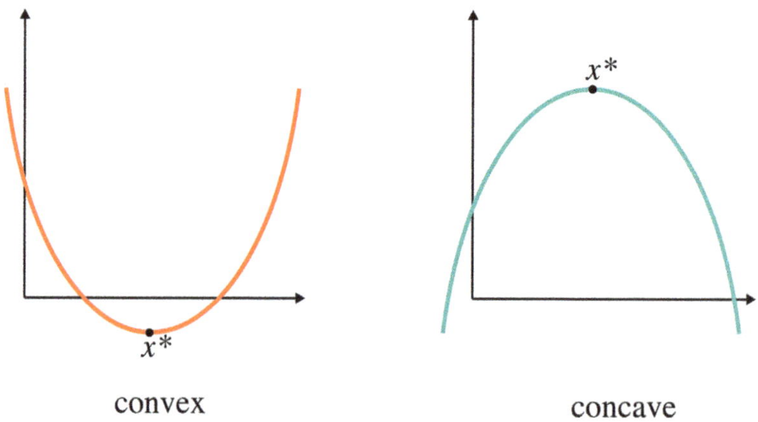

This is still not an SDP, but it can be turned into one by making use of example 2.3, in particular (2.8), which gave an SDP for the trace norm $\|A\|_1$ of an arbitrary Hermitian operator A. Here there are two additional complications: (i) whereas previously A was the *data* of the problem, here $A = \rho - \sigma$ is instead a *variable;* (ii) as stated above, we now want to *minimise* the trace norm rather than just calculate it. Luckily both of these complications are harmless. First, in (2.8) the variable X is never multiplied by the data A. Thus, although here $A = \rho - \sigma$ is now a variable, it doesn't actually stop the problem from being linear, and therefore from being an SDP.

Second, since the SDP (2.8) for the trace norm is a minimisation, and we want to consider here the *best case*—itself corresponding to *minimisation*—we end up with a double minimisation. However, double minimisations are still convex optimisation problems, and in this case our problem *remains an SDP*. Altogether, we arrive at the following SDP for finding the closest state in trace distance to a target state, given the measurement results:

$$\text{minimise} \quad \frac{1}{2}\text{tr}(X) \tag{3.5a}$$

$$\text{subject to} \quad -X \preccurlyeq \rho - \sigma \preccurlyeq X, \tag{3.5b}$$

$$\text{tr}(M_x\rho) = m_x, \qquad x = 1, \ldots, N \tag{3.5c}$$

$$\text{tr}(\rho) = 1, \tag{3.5d}$$

$$\rho \succcurlyeq 0. \tag{3.5e}$$

Incidentally, we can now understand mathematically why it is not possible to analyse the worst case, i.e. to find the state which is as far as possible from the target state. In such a case, we would end up with almost the same problem as (3.4), except the objective would now be to maximise, rather than minimise. However, we would then not be able to end up at a form like (3.5), since there it was possible to combine the two minimisations—one over ρ and one over X—into a single minimisation, and obtain an SDP. To evaluate the worst case, we would need to maximise over ρ and minimise over X, and these *cannot be combined* into a single optimisation. This reflects the fact that to find the worst case, as already realised, would require maximising a convex function, which is a fundamentally different type of problem.

3.1.2 Fidelity estimation

We now turn our attention to a second notion of closeness between states—namely the fidelity. We will start by specialising to the problem of determining how close ρ is to a *pure* quantum state $\sigma = |\psi\rangle\langle\psi|$. As will be seen, this is more straightforward than the general problem of estimating the fidelity to a *mixed* state. In particular, the fidelity between a state ρ (which can be either mixed or pure) and a pure state $\sigma = |\psi\rangle\langle\psi|$ is

$$F(\rho, \sigma) = \langle\psi|\rho|\psi\rangle = \text{tr}(|\psi\rangle\langle\psi|\rho), \tag{3.6}$$

which, notably, is a *linear* function of ρ. We can thus use this as the objective function directly, and arrive at the following SDP to find the closest state in fidelity to a target state, i.e. the state with *maximum fidelity* with the target state $\sigma = |\psi\rangle\langle\psi|$,

$$\text{maximise} \quad \text{tr}(|\psi\rangle\langle\psi|\rho) \tag{3.7a}$$

$$\text{subject to} \quad \text{tr}(M_x\rho) = m_x \qquad x = 1, \ldots, N, \tag{3.7b}$$

$$\text{tr}(\rho) = 1, \tag{3.7c}$$

$$\rho \succcurlyeq 0. \tag{3.7d}$$

This SDP again allows us to understand the *best case*, i.e. the largest fidelity among all possible states that are compatible with the observed data.

Interestingly, since the fidelity is linear, we can now consider the associated *minimisation* problem, i.e. to replace the maximisation in (3.7) by a minimisation. In this case, we then find the *worst-case* fidelity, i.e. the smallest fidelity among all possible states consistent with the measurement results.

This is somewhat surprising, since we saw for the trace distance that it isn't possible to obtain the worst-case state. The key difference here is that the fidelity to a pure state is linear, meaning it is both *convex and concave*, and so we can in fact also solve the worst-case problem too. Unfortunately, as will be seen below, this is a special property of the fidelity with pure states. In the more general case, of fidelity with a mixed target state, it is again only possible to solve the best-case scenario.

Moving on to the more general problem of estimating the fidelity between two mixed states, the general expression for the fidelity is given by

$$F(\rho, \sigma) = \left\| \sqrt{\rho}\sqrt{\sigma} \right\|_1^2, \tag{3.8a}$$

$$= \left[\mathrm{tr}\left(\sqrt{\sqrt{\sigma}\rho\sqrt{\sigma}}\right)\right]^2, \tag{3.8b}$$

which is now a non-linear function of both ρ and σ. This means that, just as with the trace distance, we can no longer directly use this as the objective function in an SDP. Although non-linear, the fidelity is nevertheless a *concave* function (in either state), meaning that

$$F(q\rho_0 + (1 - q)\rho_1, \sigma) \geqslant qF(\rho_0, \sigma) + (1 - q)F(\rho_1, \sigma), \tag{3.9}$$

for $q \in [0, 1]$, and similarly for σ. This says that the fidelity between the average state $\rho' = q\rho_0 + (1 - q)\rho_1$ and σ is never smaller than the average of the fidelities between ρ_0 and σ and ρ_1 and σ.

The fact that the fidelity is concave means—once again—that it will only be possible to estimate the *best-case* fidelity, since we will be able to consider *maximising* the fidelity (and now a larger fidelity means a closer state), given the constraints arising from the measurements result.

The difficulty that must still be overcome is to understand how we can evaluate the fidelity objective function using semidefinite programming. We will return to this topic in more detail in the 'Advanced topics' section 3.5.1, but just as with the trace distance, even though the fidelity is a non-linear function, it can nevertheless be expressed as an SDP by introducing auxiliary variables. As shown in section 3.5.1, the square-root-fidelity is in fact given by the following SDP:

$$\sqrt{F(\rho, \sigma)} = \text{maximise} \quad \mathrm{tr}(Y) \tag{3.10a}$$

$$\text{subject to} \quad \begin{pmatrix} \rho & Y + iZ \\ Y - iZ & \sigma \end{pmatrix} \succcurlyeq 0. \tag{3.10b}$$

We have written the constraint (3.10b) in a relatively intuitive way, arranging the variables Y and Z into a *block matrix*, along with the data of the problem—the quantum states ρ and σ. It is important to appreciate that this inequality constraint is equivalent to the standard form as in (2.1c), namely of the form $\Gamma(X) \preccurlyeq C$. In particular, since we have a pair of variables Y and Z, we have a map $\Gamma(Y, Z)$, which can be written

$$\Gamma(Y, Z) = |0\rangle\langle 0| \otimes \rho + |0\rangle\langle 1| \otimes (Y + iZ) + |1\rangle\langle 0| \otimes (Y - iZ) + |1\rangle\langle 1| \otimes \sigma \quad (3.11)$$

with a similar technique applicable for any block matrix. As can be seen, the block matrix form of (3.10b) is much easier to read than the form of (3.11), and hence we will use the block form whenever appropriate to do so.

Returning to the problem of estimating the largest fidelity with a target state, we can now directly use (3.10) in order to recast the problem as an SDP. In particular, the following SDP is arrived at, which evaluates the best-case (square-root) fidelity between any state ρ consistent with the data, and a target state σ:

$$\text{maximise} \quad \operatorname{tr}(Y) \quad (3.12a)$$

$$\text{subject to} \quad \begin{pmatrix} \rho & Y + iZ \\ Y - iZ & \sigma \end{pmatrix} \succcurlyeq 0, \quad (3.12b)$$

$$\operatorname{tr}(M_x\rho) = m_x \qquad x = 1, \ldots, N, \quad (3.12c)$$

$$\operatorname{tr}(\rho) = 1, \quad (3.12d)$$

$$\rho \succcurlyeq 0. \quad (3.12e)$$

Note that, compared with the SDP formulation of $\sqrt{F(\rho, \sigma)}$, we have relaxed ρ from being *input data* to being a *variable*, constrained by the measurement results. Crucially, as can be seen, this problem remains *linear*, and subsequently remains an SDP. We can also note that, just as for the trace distance, the above works since the SDP formulation of the fidelity was a *maximisation* problem. By relaxing ρ, the best-case state given the constraints, is now naturally found.

In principle, distances other than the trace distance and fidelity can also be considered in state estimation. As the above two examples have hopefully demonstrated, these distances need not be linear functions, but they must be expressible themselves as SDPs.

3.1.3 Finite statistics

We now turn our attention to a different aspect of quantum state estimation, that of *finite statistics*. In practice, due to the fact that every experiment can only handle a finite number of measurement rounds, it is rather unreasonable to assume that the expectation values $m_1, \ldots, m_N$ are known exactly. In reality, these will have been estimated themselves, and will only be known up to some level of uncertainty. Let us therefore assume that the result of each measurement is only known to lie in some interval $[m_x - \Delta_x, m_x + \Delta_x]$. The problem of interest now is the feasibility problem of determining whether there is a quantum state ρ which is consistent with all of these uncertain expectation values,

$$\text{find} \quad \rho \tag{3.13a}$$

$$\text{subject to} \quad m_x - \Delta_x \leqslant \mathrm{tr}(M_x\rho) \leqslant m_x + \Delta_x \qquad x = 1, \ldots, N, \tag{3.13b}$$

$$\mathrm{tr}(\rho) = 1, \tag{3.13c}$$

$$\rho \succcurlyeq 0, \tag{3.13d}$$

where to aid presentation we have displayed *both* inequality constraints related to the measurement intervals in a single line for each expectation value. Each equality constraint from the original problem (3.1) is thus replaced by a pair of inequality constraints. As will be seen later, this will have an effect on the *dual SDP*—which will have twice as many dual variables compared to the dual of the original problem.

From the perspective of the structure of the SDP, since there are no equality constraints relating ρ with the observed data in this problem, even if the set of measurements $\{M_x\}_x$ is tomographically complete, in general we would never expect to find a *unique* solution, unlike before. This is rather natural, as we would expect to have some residual uncertainty about the state, given the uncertainty in the measurement results, whenever they are consistent.

3.1.4 Relaxing the feasibility problem

In another direction, we return now to our original feasibility problem of quantum state estimation, as given in (3.1), and look at how it can be relaxed to an optimisation problem. There are a few of reasons for doing this. First, it will allow us to understand *quantitatively* how inconsistent a set of measurement results are, whenever there is no state that can lead to them. Second, from a numerical perspective, it is useful to be able to transform feasibility problems into optimisation problems, which are generally much more stable to solve. Finally, having an optimisation form will allow us to use duality, which we will show to be useful in later sections.

There are multiple ways in which the feasibility problem (3.1) can be relaxed to an optimisation problem, each with their relative merits. Here we will adapt the approach from the previous section on finite statistics, in order to obtain a relaxation of the problem, which is *always feasible*.

The key idea is to relax the equality constraints (3.1b)—to not demand that the observed expectation values are reproduced exactly, but allow them to be reproduced *approximately*, in a very similar fashion to the above section. Whereas in the above the relaxation was coming from the uncertainty due to finite statistics, here the logic is different. Here we seek to find a set of measurement results which *best approximates* the target results. If they can be reproduced exactly, then the original problem was feasible, and hence we obtain a quantum state ρ that can reproduce the statistics. If on the other hand the problem is infeasible, it will be necessary to perturb the results in order for them to be producible by some quantum state. The crucial observation is that *a sufficiently big perturbation of the results will always be producible by some quantum state.* This means that this relaxation will lead to an SDP which will be *feasible by construction.* Finally, by minimising the size of the perturbation, the best approximation is found.

It is interesting to note that this approach can be seem as complementary to that taken in sections 3.1.1 and 3.1.2 on trace distance and fidelity estimation. In particular, in those sections, in essence we sought to find the state which best approximated a target state. Here, on the other hand, we seek to find the set of measurement results which best approximates the target measurement results.

A simple way to achieve the above relaxation is to place a *uniform bound* on how much any single result can differ from the desired result. That is, we introduce a new variable δ, and replace the equality constraints (3.1b) with pairs of inequality constraints

$$m_x - \delta \leqslant \mathrm{tr}(M_x\rho) \leqslant m_x + \delta \qquad x = 1, \ldots, N. \tag{3.14}$$

This enforces that for all measurements, $|\mathrm{tr}(M_x\rho) - m_x| \leqslant \delta$. The reason for writing this in the form (3.14) is that in this form the constraint is manifestly linear in all of the variables. It is also worth noting that this is very similar to (3.13b), the difference being that previously the uncertainties Δ_x depended upon the measurement, and were input data to the problem, whereas here we simplify and consider only a single δ, which is now constant and a variable of the problem. Putting everything together, we arrive at the following relaxed SDP for quantum state estimation:

$$\text{minimise} \quad \delta \tag{3.15a}$$

$$\text{subject to} \quad m_x - \delta \leqslant \mathrm{tr}(M_x\rho) \leqslant m_x + \delta \qquad x = 1, \ldots, N, \tag{3.15b}$$

$$\mathrm{tr}(\rho) = 1, \tag{3.15c}$$

$$\rho \succcurlyeq 0. \tag{3.15d}$$

Any solution with $\delta = 0$ satisfies all of the constraints of the original problem (3.1). On the other hand, when $\delta^* > 0$, this signifies that the original problem (3.1) is infeasible, and there is no quantum state able to reproduce the data.

It is important to realise that δ^* provides us with quantitative information regarding how close the problem is to being feasible: if δ^* is small this means that there is a quantum state that is able to reproduce closely the measurement results; on the other hand, if δ^* is large, at least one of the measurement results needs to be significantly different from that observed, in order to be producible by some quantum state.

Exercises

3.1 By defining vectors $\vec{m}$ with components m_x and $\vec{m}'$ with components $m_x' = \text{tr}(M_x\rho)$, show that the SDP (3.15) can also be written as

$$\begin{aligned} \text{minimise} \quad & \|\vec{m}' - \vec{m}\|_\infty \\ \text{subject to} \quad & m_x' = \text{tr}(M_x\rho) \qquad x = 1, \ldots, N, \\ & \text{tr}(\rho) = 1, \\ & \rho \succcurlyeq 0. \end{aligned}$$

This shows that this relaxation can be understood as minimising the ℓ_∞ distance between the target data $\vec{m}$, and any set of data that can be produced in quantum mechanics.

3.2 Write down the SDP relaxation of (3.1) which minimises instead the ℓ_1 distance $\frac{1}{2}\|\vec{m} - \vec{m}'\|_1$, between the target data $\vec{m}$, and data that can be produced in quantum mechanics $\vec{m}'$.

3.1.5 Certificate of infeasibility

We now consider the situation where we are given some experimental data which is inconsistent, such that there is no quantum state that could possibly lead to this data. An interesting question is: can we certify that this is the case without having to numerically solve the SDP? Is there an *analytic* certificate that can be used to prove unequivocally that the SDP is infeasible? In this section it will be seen that duality allows us to provide such a simple *certificate*, which will convince us whenever a set of measurement results is inconsistent. This idea is rather general, and can be widely applied to guarantee that an given SDP is infeasible.

Our starting point will be the relaxed problem (3.15). Recall that we attempt to minimise δ, and if we are able to find a solution $\delta = 0$, then the associated state ρ will be consistent with the measurement results. On the other hand, if $\delta^* > 0$, then the results are inconsistent. In what follows we will show how duality can be used to guarantee that $\delta^* > 0$ without having to solve the SDP. Recall that for a minimisation problem, the values that the dual objective function can take always lower bound the optimal value of the primal problem (weak duality). Therefore, if we can find a set of dual variables such that the value of the dual objective function is strictly positive, then *with certainty* the primal problem has $\delta^* > 0$. It is precisely a collection of dual variables that lead to a positive value of the dual objective function that will constitute our certificate. Let us now put this into practice.

By introducing scalar dual variables u_x and v_x associated to the first and second inequality constraint respectively in (3.15b), a scalar z associated to (3.15c) and an operator W associated to (3.15d), the associated Lagrangian of the problem is

$$\mathcal{L} = \delta - \sum_{x=1}^{N} u_x[\text{tr}(M_x\rho) - m_x + \delta] - \sum_{x=1}^{N} v_x[m_x + \delta - \text{tr}(M_x\rho)] + z[1 - \text{tr}(\rho)] - \text{tr}(W\rho), \tag{3.16a}$$

$$= \delta\left[1 - \sum_{x=1}^{N}(u_x + v_x)\right] + \text{tr}\left[\rho\left(\sum_{x=1}^{N}(v_x - u_x)M_x - z\mathbb{1} - W\right)\right] + z + \sum_{x=1}^{N}(u_x - v_x)m_x. \tag{3.16b}$$

Note that since the primal problem is a minimisation problem, the Lagrangian is constructed to be smaller than the value of the primal objective function for all primal feasible variables, and hence we take $u_x \geqslant 0$, $v_x \geqslant 0$ and $W \geqslant 0$. Recall that this is the reason for the additional minus signs in the Lagrangian in the second, third and last term, which arise whenever we construct the Lagrangian for a minimisation problem, as shown in exercise 2.11. We can additionally make the Lagrangian independent of the primal variables by ensuring the first and second brackets vanish in (3.16b). Maximising, to obtain the best lower bound, and solving for W, which is seen to play the role of a slack variable, we arrive at the dual formulation

$$\text{maximise} \quad z + \sum_{x=1}^{N}(u_x - v_x)m_x \tag{3.17a}$$

$$\text{subject to} \quad z\mathbb{1} + \sum_{x=1}^{N}(u_x - v_x)M_x \preccurlyeq 0, \tag{3.17b}$$

$$\sum_{x=1}^{N}(u_x + v_x) = 1, \tag{3.17c}$$

$$u_x \geqslant 0 \quad v_x \geqslant 0 \quad x = 1, \ldots, N. \tag{3.17d}$$

Let us now analyse this dual a little. The first thing to notice is that the constraints (3.17c) and (3.17d), together with the fact that everywhere else only the combination $u_x - v_x$ appears, shows that the problem can be interpreted as one of optimising over a single dual variable $\vec{t} = \vec{u} - \vec{v}$, with components $t_x = u_x - v_x$, such that $\|\vec{t}\|_1 \leqslant 1$. This can be seen from the definition of the ℓ_1 unit ball from (1.62). Thus, we can express (3.17) in a simpler form:

$$\text{maximise} \quad z + \vec{t} \cdot \vec{m} \tag{3.18a}$$

$$\text{subject to} \quad z\mathbb{1} + \vec{t} \cdot \vec{M} \preccurlyeq 0, \tag{3.18b}$$

$$\|\vec{t}\|_1 \leqslant 1, \tag{3.18c}$$

where we have introduced an *operator vector* $\vec{M} = (M_1, \ldots, M_N)$, the components of which are the observables, and, as above, $\vec{m} = (m_1, \ldots, m_N)$ is the vector whose components are the measurement results.

How does this serve as a certificate for the fact that the measurement results $\vec{m}$ were inconsistent? Let us assume that we have found a set of dual feasible points—i.e. a set of dual variables $(z, \vec{t})$ satisfying the constraints—such that $\beta = z + \vec{t} \cdot \vec{m} > 0$. Now, consider the constraint (3.18b), and multiply both sides by an arbitrary quantum state ρ and take the trace. We see that

$$\mathrm{tr}\left[\rho\left(z\mathbb{1} + \vec{t} \cdot \vec{M}\right)\right] = z + \sum_{x=1}^{N} t_x(\rho M_x) \leqslant 0. \tag{3.19}$$

The first equality follows from the linearity of the trace, while the second follows from the fact that for two operators $A \succcurlyeq 0$ and $B \preccurlyeq 0$, we have $\mathrm{tr}(AB) \leqslant 0$. What this shows is that the operator inequality (3.18b) guarantees that any set of measurement results $\vec{m}' = (\mathrm{tr}(\rho M_1), \ldots, \mathrm{tr}(\rho M_N))$ that can arise by performing measurements on a quantum state, will never lead to a positive value of β. This shows that the claim that the purported results $\vec{m}$ were observed, must have been false. The dual problem is thus seen to look for a carefully chosen set of dual variables $(z, \vec{t})$ such that on the one hand, the purported results lead to a positive value of the dual objective function, while simultaneously demanding that all valid measurement results lead to a negative value. In the next section we will explore the geometry of this, but as a preview, what we have just observed can be thought of as a *separation* between what is allowed and what is not allowed, and is a common geometrical feature of certificates of this type.

Why is this better than solving the primal SDP in (3.1)? There are a number of advantages. First, given the dual variables, no optimisation is required in order to check whether the following two conditions hold: (i) $\beta = z + \vec{t} \cdot \vec{m} > 0$, and (ii) all of the eigenvalues of $W = z\mathbb{1} + \vec{t} \cdot \vec{M}$ are negative. Second, this certificate can in principle even be checked analytically, that is, we can analytically evaluate β and find the eigenvalues of W, and thus verify analytically that all of the constraints of the dual are satisfied. This is in contrast to only having a numerical solution for the primal problem. Finally, although this certificate is designed for a given set of inconsistent data $\vec{m}$, it can be used more universally, to check the inconsistency of *any* data. In particular, although it may fail to identify inconsistent data, if any set of data leads to a strictly positive value of β, then it is guaranteed to be inconsistent. We exemplify this in the following example:

Example 3.1 Equatorial plane of the Bloch sphere

In this example, we will show how the above dual SDP (3.18) can in fact be used as a method to derive the equator of the Bloch sphere. That is, any qubit, written in the form $\rho = (\mathbb{1} + \vec{r} \cdot \vec{\sigma})/2$, with $\vec{\sigma} = (\sigma_x, \sigma_y, \sigma_z)$ the vector of Pauli operators and $\vec{r}$ the Bloch vector with components $r_i = \mathrm{tr}(\sigma_i \rho)$, will be a valid density operator only if $r_x^2 + r_y^2 \leqslant 1$.

In order to obtain this result, assume that two Pauli measurements are performed, $\vec{M} = (\sigma_x, \sigma_y)$. We will use (3.18) to find the allowed measurement results $\vec{m} = (m_x, m_y)$ that can arise—and in so doing re-derive the equatorial plane of the Bloch sphere.

From (3.18b), we are seeking dual variables that satisfy

$$z\mathbb{1} + t_x\sigma_x + t_z\sigma_z \preccurlyeq 0 \tag{3.20}$$

and from (3.18c), we furthermore must have $\|\vec{t}\|_1 = |t_x| + |t_z| \leqslant 1$. Since any operator of the form $n_x\sigma_x + n_y\sigma_y$ has eigenvalues ± 1 whenever $\|\vec{n}\|_2 = \sqrt{n_x^2 + n_y^2} = 1$, we see that the eigenvalues of the operator $z\mathbb{1} + t_x\sigma_x + t_y\sigma_y$ are

$$\lambda_\pm = z \pm \|\vec{t}\|_2, \tag{3.21}$$

and so we will satisfy (3.20) whenever

$$z + \|\vec{t}\|_2 \leqslant 0. \tag{3.22}$$

We can therefore take $z = -\|\vec{t}\|_2$, so that the inequality is saturated. With this choice of z, the dual objective function is

$$-\|\vec{t}\|_2 + \vec{t} \cdot \vec{m}. \tag{3.23}$$

Now, any set of measurement outcomes $\vec{m} = (m_x, m_y)$ for which this is strictly positive is inconsistent, since we have just arranged it such that this can never happen. That is, $\vec{m}$ is inconsistent if

$$\vec{t} \cdot \vec{m} > \|\vec{t}\|_2. \tag{3.24}$$

Up to this point, we still haven't specified $\vec{t}$, other than the requirement that $\|\vec{t}\|_1 \leqslant 1$. Let us therefore choose

$$\vec{t} = \frac{\vec{m}}{\|\vec{m}\|_1}. \tag{3.25}$$

This implies that $\|\vec{t}\|_2 = \|\vec{m}\|_2 / \|\vec{m}\|_1$, and so the data is inconsistent whenever

$$\|\vec{m}\|_2^2 > \|\vec{m}\|_2, \tag{3.26}$$

i.e. $\|\vec{m}\|_2 > 1$. In other words, consistent measurement outcomes must satisfy $\|\vec{m}\|_2 \leqslant 1$. Written out in full, this equates to $m_x{}^2 + m_y{}^2 \leqslant 1$. Since $\vec{m}$ is nothing but the components of the Bloch vector of a qubit in the equatorial plane, this re-derives the equatorial plane of the Bloch sphere.

Of course this result is well-known and can be obtained more directly, by finding the eigenvalues of a qubit density operator—we did not need to resort to the duality theory of SDPs to derive this result. However, the utility of going through this example is that in more complicated situations, involving for example non-Pauli measurements, or higher dimensions, this approach can still be applied, and provides a general method that can be used to understand the limitations of the quantum state space, and the measurement statistics it leads to.

As an exercise below, this same procedure is carried out with all three Pauli measurements, from which the entire Bloch sphere of a qubit can similarly be recovered.

Exercises

3.3 Apply the same argument as in exercise 3.1, but for data $\vec{m} = (m_x, m_y, m_z)$ arising from the three Pauli measurements $\vec{M} = (\sigma_x, \sigma_y, \sigma_z)$, to show that the data is inconsistent whenever $\|\vec{m}\|_2 > 1$. Explain why this re-derives the Bloch sphere.

3.1.6 Geometrical interpretation

We finish our exploration of quantum state estimation by investigating how the above certificates of infeasibility can be understood geometrically. Understanding

SDPs and their duality geometrically can be very powerful, and provides insight and intuition for how and why duality works.

Our starting point is to think geometrically about the *space of quantum states* and *space of measurement data*. The space of all quantum states of a fixed, finite dimension d is the set

$$\mathcal{S} = \{\rho \mid \rho \succcurlyeq 0, \operatorname{tr}(\rho) = 1\}, \tag{3.27}$$

where ρ is an operator acting on $\mathbb{C}^d$. This is a convex set, which we can always visualise as a subset of $\mathbb{R}^{d^2}$. For a fixed set of measurements $\vec{M}$, the set of all quantum states $\mathcal{S}$ gets mapped into a *space of measurement results*, which is the image of $\mathcal{S}$ under the mapping

$$\rho \mapsto \vec{m} = (\operatorname{tr}(M_1\rho), \ldots, \operatorname{tr}(M_N\rho)) \tag{3.28}$$

which can be represented as a subset of $\mathbb{R}^N$. We denote this image by $\mathcal{M}$. This mapping is illustrated in figure 3.1. Because this is a linear map, the convexity of the state space is preserved, and the space of all measurement results is also a convex set—see exercise 3.4.

We can understand many features of the state estimation problem from this geometrical perspective. First, note that the estimation problem is an *inverse problem*: we are given a point in the measurement space, and ask for the quantum state(s) ρ under the inverse mapping. When the set of measurements is not tomographically complete, then multiple states lead to the same set of measurement results. This shows that there is a *loss of information*. As such, there is not a unique solution to the inverse problem. In this case, the dimension of the measurement space will be smaller than the dimension of the state space. On the other hand, when

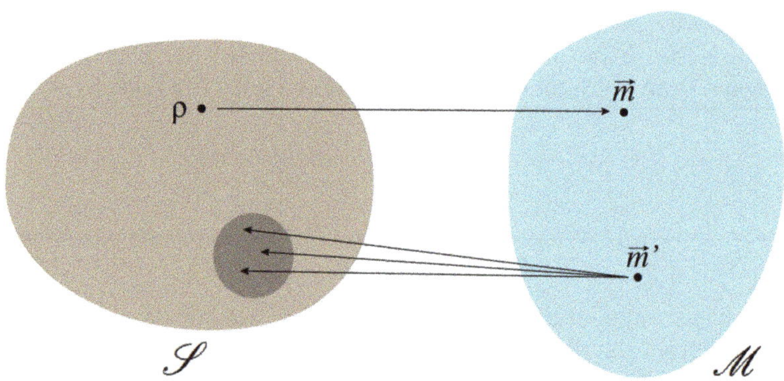

Figure 3.1. *Mapping between state space and space of measurement results of a fixed set of measurements.* An illustration of the mapping of the quantum state space $\mathcal{S}$ into the set of measurement results $\mathcal{M}$, under the mapping $\rho \mapsto \vec{m} = (\operatorname{tr}(M_1\rho), \ldots, \operatorname{tr}(M_N\rho))$. For illustrative purposes both spaces are depicted as being 2-D, although in general they are both high-dimensional bodies of different dimension. Each quantum state ρ is mapped to a single vector of measurement results $\vec{m}$. In general, the mapping can be many-to-one, with multiple quantum states being mapped to the same point in $\mathcal{M}$. This means that the inverse mapping, from a vector $\vec{m}'$ leads to a subset of the state space $\mathcal{S}$, as depicted.

the measurements are tomographically complete, there is no loss of information and a one-to-one mapping between the two spaces. This implies that the dimension of the measurement space must be equal or greater to the dimension of the state space.

Most interestingly, we can gain insight into the certificate of infeasibility. Geometrically, if a fictitious set of measurement results $\vec{m}_{\text{inc}}$ is inconsistent, this means it *lies outside of the image of* $\mathcal{S}$. A remarkable fact from convex geometry now comes into play: *it is always possible to find a hyperplane which separates a point from a convex set*. This is somewhat obvious in one, two and three dimensions, and in fact it holds in all dimensions. The certificate of infeasibility in *precisely the specification of such a separating hyperplane*, as depicted in figure 3.2.

As we will see in many instances throughout this book, dual SDPs can often be interpreted as providing *certificates* or *witnesses* of certain properties, and these can be understood geometrically as specifying separating hyperplanes.

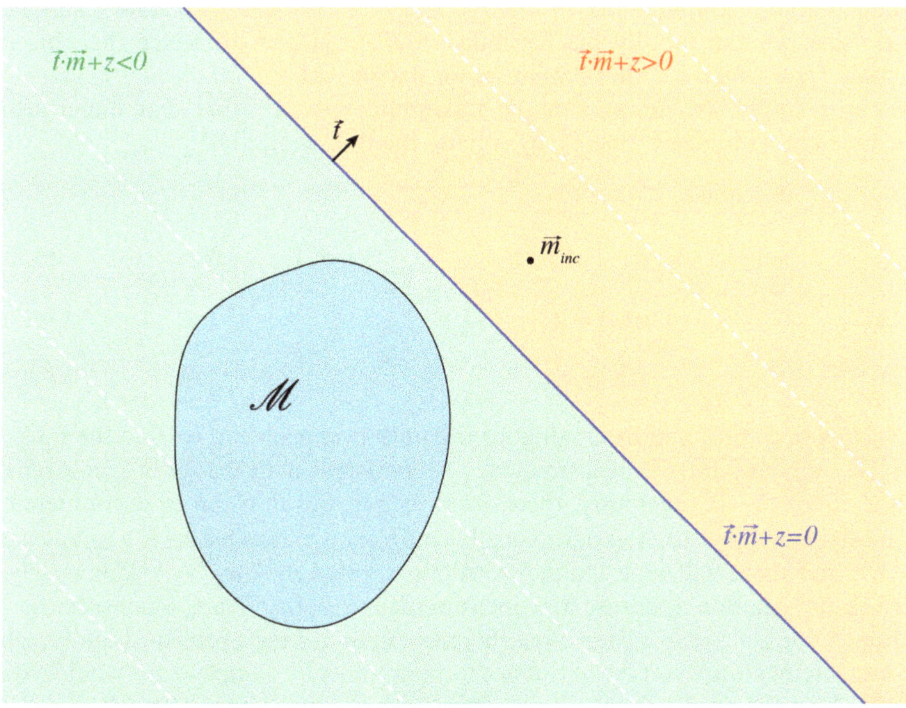

Figure 3.2. *Geometric interpretation of certificate of infeasibility.* An illustrative example in 2-D of how the certificate of infeasibility can be viewed as a *separating hyperplane* where $z + \vec{t} \cdot \vec{m} = 0$, which separates $\mathcal{M}$—the image of $\mathcal{S}$ (blue set)—from the supposed set of measurement results $\vec{m}_{\text{inc}}$. The vector $\vec{t}$ specifies a direction, and z specifies an offset, such that the image of $\mathcal{S}$ lies in the (green) region where $z + \vec{t} \cdot \vec{m} < 0$ and $\vec{m}_{\text{inc}}$ lies in the orange region, where $z + \vec{t} \cdot \vec{m} > 0$. Re-arranging, we can view $-z$ as the 'bound' on the linear function $\vec{t} \cdot \vec{m}$ that separates the two regions.

Exercises

3.4 Show that the set $\mathcal{M}$ of measurement results—the image of $\mathcal{S}$ from (3.27) under the mapping (3.28)—is a convex set. That is, show that if two sets of results $\vec{m}_1$ and $\vec{m}_2$ are both in $\mathcal{M}$, then so is any set of results of the form $\vec{m}' = p\vec{m}_1 + (1 - p)\vec{m}_2$ for $0 \leqslant p \leqslant 1$.

3.1.7 Property estimation

It can sometimes be the case that we are less interested in estimating the full state ρ that led to the results $\vec{m}$, and more interested in knowing some property of it. If this property can be calculated by a linear function $f(\rho)$, such as energy or magnetisation, then we can find the range of values which are possible, by finding both the minimum and maximum value of this property that any quantum state ρ can have. To do this, we can modify the feasibility SDP (3.1) and introduce the objective function $f(\rho)$, which is either minimised or maximised.

As an example, we can estimate the maximum expected value of an unperformed measurement with observable $\tilde{M}$ by solving the following SDP

$$\text{maximise} \quad \text{tr}(\tilde{M}\rho) \tag{3.29a}$$

$$\text{subject to} \quad \text{tr}(M_x\rho) = m_x \qquad x = 1, \ldots, N, \tag{3.29b}$$

$$\text{tr}(\rho) = 1, \tag{3.29c}$$

$$\rho \succcurlyeq 0. \tag{3.29d}$$

By solving both this, and the analogous minimisation problem, we find the range of possible outcomes for the measurement $\tilde{M}$, consistent with the set of measurement results observed. As previously, there are three regimes that can be encountered. If the measurements $\vec{M}$ are tomographically complete, the feasible set is a unique state $\mathcal{F} = \rho^*$, and there will be a unique expectation value $\tilde{m} = \text{tr}(\tilde{M}\rho^*)$. Conversely, if there is no solution (because the measurement results $\vec{m}$ are inconsistent), the optimal value is $\alpha^* = -\infty$, signalling the infeasibility of the problem. Finally, when the measurements are consistent but not tomographically complete, we would expect there to be a range of values, all consistent with the measurement results.

In sections 3.1.1 and 3.1.2 we saw that it is possible to calculate the trace distance and fidelity to a target state σ. From the current perspective, we can view these as two further instances of property testing—of the state having the property of being σ. As was seen previously, even though the trace distance and fidelity are non-linear functions, both of these properties can nevertheless be estimated using semidefinite programming. It is possible to estimate other properties—specified by non-linear functions—using similar techniques.

3.2 The quantum marginal problem

In this final section we will now consider the quantum marginal problem. This problem is closely related to quantum state estimation, with the key difference being in the type of information given. In the previous sections of this chapter, we considered being given a set of measurement results, and were interested in estimating the quantum state—or properties of it—from this measurement data. We discussed in those sections the idea of tomographically complete data, i.e. data which is sufficient to uniquely determine the quantum state.

In the context of the quantum marginal problem, we will be interested in *multipartite* quantum states, comprising a number of subsystems. The assumption made now is that we have perfect knowledge of some of the subsystems—i.e. some of the *marginal* states, more commonly referred to as reduced density operators. This perfect knowledge can be thought of as being obtained by having tomographically complete data for these subsystems—i.e. by having performed a tomographically complete set of measurements, and having the corresponding measurement results. The basic problems of the quantum marginal problem are then to either determine the *global state* consistent with all of the marginals, or to estimate some property of this global state, given access only to the marginals.

We could now phrase everything just as was done above, in terms of measurement results. However, since the focus is now exclusively on situations where we have full knowledge of a number of subsystems, it is useful to adopt a more direct approach, and simply specify the reduced density operators of those subsystems. On the one hand, this is a conceptually cleaner way to approach the problem. On the other hand, it is an important lesson to realise that even in this formulation, we can still apply the techniques of semidefinite programming.

As a concrete example, let us consider the direct quantum analogue of the (classical) marginal problem considered in example 1.1. Therefore, consider a tripartite quantum system, with subsystems labelled by X, Y and Z. Suppose that we have knowledge of the bipartite reduced density operators ρ_{XY}, ρ_{XZ} and ρ_{YZ}, but not of the joint state ρ_{XYZ}. Consider that we want to determine whether there is any joint state consistent with the reduced density operators, and to find an example of one if it does. This can be cast as the following feasibility SDP:

$$\text{find} \quad \sigma_{XYZ} \tag{3.30a}$$

$$\text{subject to} \quad \mathrm{tr}_Z(\sigma_{XYZ}) = \rho_{XY}, \tag{3.30b}$$

$$\mathrm{tr}_Y(\sigma_{XYZ}) = \rho_{XZ}, \tag{3.30c}$$

$$\mathrm{tr}_X(\sigma_{XYZ}) = \rho_{YZ}, \tag{3.30d}$$

$$\sigma_{XYZ} \succcurlyeq 0, \quad \mathrm{tr}(\sigma_{XYZ}) = 1. \tag{3.30e}$$

As with the state estimation problem, there are a number of interesting variants of this basic problem that can be considered. First, we can consider both trace distance and fidelity estimation of the global state (or a marginal state) to a target state, in analogy to the problems studied in sections 3.1.1 and 3.1.2. We study these generalisations in exercises 3.5 and 3.6 respectively. Second, we can also consider the analogous problem of property estimation from section 3.1.7. This is studied in exercise 3.7 below.

We can also consider a problem which is closely related to the finite-statistics version of state estimation from section 3.1.3. Here we imagine that instead of knowing the bipartite reduced density operators exactly, they are only known approximately. One way to model this is to assume that the marginals need to be ϵ-close to a fixed state, i.e. to demand

$$\|\sigma_{XY} - \rho_{XY}\|_1 \leqslant \epsilon, \tag{3.31}$$

where we have used the shorthand $\sigma_{XY} = \mathrm{tr}_Z(\sigma_{XYZ})$, and similarly for σ_{XZ} and σ_{YZ}. We note that, just as when considering finite statistics, ϵ is considered as being data which is specified in the problem—allowing us to impose either stricter or weaker constraints on how close the reduced density operators of ρ_{XYZ} need to be to the target states. Thus, we arrive at the following optimisation problem

$$\text{find} \quad \sigma_{XYZ} \tag{3.32a}$$

$$\text{subject to} \quad \|\sigma_{XY} - \rho_{XY}\|_1 \leqslant \epsilon, \tag{3.32b}$$

$$\|\sigma_{XZ} - \rho_{XZ}\|_1 \leqslant \epsilon, \tag{3.32c}$$

$$\|\sigma_{YZ} - \rho_{YZ}\|_1 \leqslant \epsilon, \tag{3.32d}$$

$$\sigma_{XYZ} \succcurlyeq 0, \quad \mathrm{tr}(\sigma_{XYZ}) = 1. \tag{3.32e}$$

As written, this is not an SDP, due to the non-linear inequality constraints (3.32b)–(3.32d). However, we can make use of the SDP characterisation of the ϵ trace norm ball, as given in exercise 2.17, itself a small extension of the unit trace norm ball from equation 2.36. In particular, we arrive at

$$\text{find} \quad \sigma_{XYZ} \tag{3.33a}$$

$$\text{s.t.} \quad \sigma_{XY} - \rho_{XY} = \omega_{XY} - \zeta_{XY}, \quad \mathrm{tr}(\omega_{XY} + \zeta_{XY}) = \epsilon, \quad \omega_{XY} \succcurlyeq 0, \; \zeta_{XY} \succcurlyeq 0, \tag{3.33b}$$

$$\sigma_{XZ} - \rho_{XZ} = \omega_{XZ} - \zeta_{XZ}, \quad \mathrm{tr}(\omega_{XZ} + \zeta_{XZ}) = \epsilon, \quad \omega_{XZ} \succcurlyeq 0, \; \zeta_{XZ} \succcurlyeq 0, \tag{3.33c}$$

$$\sigma_{YZ} - \rho_{YZ} = \omega_{YZ} - \zeta_{YZ}, \quad \mathrm{tr}(\omega_{YZ} + \zeta_{YZ}) = \epsilon, \quad \omega_{YZ} \succcurlyeq 0, \; \zeta_{YZ} \succcurlyeq 0, \tag{3.33d}$$

$$\sigma_{XYZ} \succcurlyeq 0, \quad \mathrm{tr}(\sigma_{XYZ}) = 1. \tag{3.33e}$$

In order to emphasise the relationship between this formulation and (3.32), we have presented four constraints per line, which collectively replace the corresponding trace norm constraint from the former. Note also that we have had to introduce two new variables per constraint, thus arriving at a larger problem. Nevertheless, this form is explicitly an SDP, and shows that it is possible to solve the analogue of the finite statistics problem for the quantum marginal problem. Finally, although we chose to place a constraint in terms of the trace norm, it is also possible in principle impose any distance-type constraint, as long as it can be expressed as an SDP. Examples of such include other norms, such as the operator norm, as well as the fidelity, both of which are left as exercises below.

Finally, we can also consider turning the problem from a feasibility SDP into a (standard) optimisation SDP. As always, there are numerous approaches that one can take to achieve this. Here we will consider a particularly simple approach which can be used when the target marginal states are *pure*. In this case, a relaxation of the marginal problem is to find a global state whose marginals have the largest average fidelity with the target pure marginals.

As a particular example, let us consider the analogue situation from exercise 1.9(c), where only two of the marginals are specified. Let us therefore assume that $\rho_{XY} = |\psi_{XY}\rangle\langle\psi_{XY}|$ and $\rho_{YZ} = |\psi_{YZ}\rangle\langle\psi_{YZ}|$. The largest average fidelity that can be achieved with these two states is given by the following simple SDP

$$\text{maximise} \quad \frac{1}{2}(\langle\psi_{XY}|\sigma_{XY}|\psi_{XY}\rangle + \langle\psi_{YZ}|\sigma_{YZ}|\psi_{YZ}\rangle) \tag{3.34a}$$

$$\text{subject to} \quad \sigma_{XYZ} \succcurlyeq 0, \quad \text{tr}(\sigma_{XYZ}) = 1. \tag{3.34b}$$

In exercise 1.9(c) it was shown that for the *classical* version of this problem, with only two pairwise marginals specified, as long as they were compatible—agreeing on the marginal distribution of the common random variable—then a global distribution always exists which perfectly reproduces the two marginals. In the quantum setting, interestingly, this is no longer the case. In particular, given two states $|\psi_{XY}\rangle$ and $|\psi_{YZ}\rangle$ such that $\rho_Y = \text{tr}_X(|\psi_{XY}\rangle\langle\psi_{XY}|) = \text{tr}_Z(|\psi_{YZ}\rangle\langle\psi_{YZ}|)$, in general there *will not* be a joint state of the three particles. The origin of this is *monogamy of entanglement*—a particle cannot simultaneously be highly entangled with two other particles, which introduces completely novel constraints into the marginal problem. As will be seen below in exercise 3.8, we can use the dual of (3.34) to show that a particle cannot simultaneously be in a pure entangled state with two other particles.

In chapter 5 we will return to an important variant of the marginal problem when studying the notion of a *k-symmetric extension*, a powerful tool in the theory of entanglement.

Exercises

3.5 In this exercise we will consider problems involving finding the closest state to a target state—either the global state, or a marginal—in terms of the trace distance.

(a) Consider a situation involving three particles, where we are given all of the pairwise marginals, ρ_{XY}, ρ_{XZ} and ρ_{YZ}, and wish to estimate the trace distance between the closest compatible state σ_{XYZ} and a target global state ρ_{XYZ}.

Write down the optimisation problem analogous to (3.4) that needs to be solved. *This problem will not be an SDP.*

(b) Make use of (2.8) in order to re-express the optimisation problem from part (a) as an SDP.

(c) Repeat the exercise from parts (a) and (b), assuming now instead that only the marginal states ρ_{XY} and ρ_{XZ} are specified, and that the goal is to estimate the trace distance between the closest compatible marginal state σ_{YZ} and a target marginal state ρ_{YZ}.

3.6 In this exercise we will repeat the same calculations as in the previous exercise, except instead of optimising the trace distance, we now optimise the fidelity.

(a) Consider a situation involving three particles, where we are given all of the pairwise marginals, ρ_{XY}, ρ_{XZ} and ρ_{YZ}, and wish to estimate the fidelity between the closest compatible state σ_{XYZ} and a target global state ρ_{XYZ}. Write down the optimisation problem that needs to be solved. *This problem will not be an SDP.*

(b) Make use of (3.10) in order to re-express the optimisation problem from part (a) as an SDP.

(c) Repeat the exercise from parts (a) and (b), assuming now instead that only the marginal states ρ_{XY} and ρ_{XZ} are specified, and that the goal is to estimate the fidelity between the closest compatible marginal state σ_{YZ} and a target marginal state ρ_{YZ}.

3.7 In this exercise we will consider *property estimation* in the context of the marginal problem. Consider that we are given the Hamiltonian H of three particles, and the three pairwise marginal states ρ_{XY}, ρ_{XZ} and ρ_{YZ}. Our goal is to find the range of average energies that the system can have, consistent with these marginals.

(a) Write down an SDP that evaluates the minimum possible average energy any global state σ_{XYZ} can have, given the Hamiltonian H and marginal states.

(b) Write down an SDP that evaluates the maximum possible average energy any global state σ_{XYZ} can have, given the Hamiltonian H and marginal states.

3.8 In this exercise we will derive the dual SDP of (3.34) and use it to show that, when in a pure state, a qubit cannot simultaneously be entangled with two other qubits.

(a) Write down the Lagrangian associated to the primal SDP (3.34), using M_{XYZ} and μ as dual variables for the first and second constraint, respectively.

(b) Identify the constraints that need to be satisfied by the dual variables in order that

(i) The Lagrangian upper bounds the value of the primal objective function for all primal feasible variables.

(ii) The Lagrangian becomes independent of the primal variables.

(c) Use parts (a) and (b) to show that, after solving for the slack variables, the dual SDP to (3.34) is

$$\text{minimise} \quad \mu \tag{3.35a}$$

$$\text{subject to} \quad \mu\mathbb{1} \geqslant \frac{1}{2}(|\psi_{XY}\rangle\langle\psi_{XY}| \otimes \mathbb{1}_Z + \mathbb{1}_X \otimes |\psi_{YZ}\rangle\langle\psi_{YZ}|). \tag{3.35b}$$

This SDP can be solved explicitly, with optimal value given by

$$\frac{1}{2}\ \| |\psi_{XY}\rangle\langle\psi_{XY}| \otimes \mathbb{1}_Z + \mathbb{1}_X \otimes |\psi_{YZ}\rangle\langle\psi_{YZ}| \|_\infty .$$

(d) For two projection operators Π_1 and Π_2, the following bound can be shown to hold:

$$\|\Pi_1 + \Pi_2\|_\infty \leqslant 1 + \sqrt{\|\Pi_2\Pi_1\Pi_2\|_\infty}\,. \tag{3.36}$$

Writing the states $|\Psi_{XY}\rangle$ and $|\Psi_{YZ}\rangle$ in terms of their Schmidt decompositions,

$$|\Psi_{XY}\rangle = \sum_i \sqrt{p_i}|i_X\rangle|i_Y\rangle, \qquad |\Psi_{YZ}\rangle = \sum_i \sqrt{q_i}|i_Y\rangle|i_Z\rangle,$$

use (3.36) to show that the optimum value of the dual SDP (3.35) is not larger than

$$\mu^* \leqslant \frac{1}{2}\left(1 + \max_i \sqrt{p_i q_i}\right). \tag{3.37}$$

(e) Explain why the answer to part (d) shows that a qubit cannot simultaneously be in a pure entangled state with two other particles.

3.3 Concluding remarks

In this chapter we began our exploration of applying semidefinite programming techniques to concrete problems in quantum information science. We mainly focused on the problem of quantum state estimation, which allowed us to study a variety of interesting features of SDPs. In particular, the main results described in this chapter are:

- **Quantum state estimation.** The first problem dealt with was determining the existence of a quantum state compatible with a set of measurement data. We introduced a few variants of this problem, in particular, focusing on different figures of merit, such as the trace distance and the fidelity. This demonstrated that problems involving non-linear objective functions can be cast as SDPs.
- **Double minimisation.** We showed that it is possible to cast problems involving double minimisation or double maximisation as SDPs—see (3.5).
- **Certificates of infeasibility.** When considering feasibility problems reformulated as optimisation problems, the dual variables of SDPs can often be viewed as providing certificates of infeasibility. This follows from the facts that (i) every feasibility SDP can be turned into an optimisation SDP in which a positive solution indicates that the original problem was infeasible; (ii) due to weak duality, the value of the objective function of the dual SDP lower bounds the optimal value of the primal SDP. This allows us to analytically certify infeasibility.

3.4 Further reading

- Nielsen M A and Chuang I L 2000 *Quantum Computation and Quantum Information* (Cambridge: Cambridge University Press) http://doi.org/10.1017/CBO9780511976667

3.5 Advanced topics

3.5.1 The fidelity SDP

In this section we will derive the SDP formulation for fidelity as given in (3.10). Recall that for a general pair of quantum states ρ and σ, the fidelity is defined by

$$F(\rho, \sigma) = \|\sqrt{\rho}\sqrt{\sigma}\|_1^2. \tag{3.38}$$

Our starting point in this section will be *Uhlmann's theorem*, which gives an alternative expression for the fidelity in terms of *purifications* of ρ and σ. Recall that a purification of a (mixed) state ω is a bipartite state $|\chi\rangle$ such that

$$\mathrm{tr}_B[|\chi\rangle\langle\chi|] = \omega, \tag{3.39}$$

where we label the systems as A and B.

Uhlmann's theorem then gives a classification of the fidelity in terms of the biggest overlap between any purification of ρ and σ, namely

$$F(\rho, \sigma) = \text{maximise} \quad |\langle\psi|\phi\rangle|^2 \tag{3.40a}$$

$$\text{subject to} \quad \mathrm{tr}_B\,[|\psi\rangle\langle\psi|] = \rho, \tag{3.40b}$$

$$\mathrm{tr}_B[|\phi\rangle\langle\phi|] = \sigma. \tag{3.40c}$$

Let us consider that we have found an arbitrary purification of ρ and an arbitrary purification of σ, which we denote by $|\psi\rangle$ and $|\phi\rangle$ respectively. Consider then the following state

$$|\Psi\rangle = \frac{1}{\sqrt{2}}(|\psi\rangle|0\rangle + e^{i\theta}|\phi\rangle|1\rangle), \tag{3.41}$$

where we have now introduced a third system, labelled C, taken to be a qubit, and an arbitrary phase factor $e^{i\theta}$ which we will fix below.

The basic observation we make is that if $|\psi\rangle$ and $|\phi\rangle$ are *similar*—i.e. if they are close to being the same state, or have a large overlap $|\langle\psi|\phi\rangle|$—then the state $|\Psi\rangle$ *cannot be very entangled.*

In particular, as will be seen later in chapter 5, the entanglement of a pure state can be determined by the reduced density operator of one subsystem. A state is entangled if and only if the reduced density operator (on any subsystem) is mixed. In our case, the reduced density operator of C is

$$\omega_C = \mathrm{tr}_{AB}[|\Psi\rangle\langle\Psi|] = \frac{1}{2}(|0\rangle\langle 0| + |1\rangle\langle 1| + e^{-i\theta}\langle\phi|\psi\rangle|0\rangle\langle 1| + e^{i\theta}\langle\psi|\phi\rangle|1\rangle\langle 0|). \tag{3.42}$$

Let us now choose the angle θ such that $e^{i\theta}\langle\psi|\phi\rangle$ is real and nonnegative (and therefore equal to $e^{-i\theta}\langle\phi|\psi\rangle$ and, moreover $|\langle\psi|\phi\rangle|$). The purity of ω_C is then seen to be

$$\mathrm{tr}[\omega_C^2] = \frac{1}{2}(1 + |\langle\psi|\phi\rangle|^2). \tag{3.43}$$

This shows that when $|\psi\rangle$ and $|\phi\rangle$ are orthogonal, such that $|\langle\psi|\phi\rangle| = 0$, then the purity of ω_C is minimal, since $\omega_C = \frac{1}{2}\mathbb{I}$ is the maximally mixed state in this case. On the other hand, when $|\langle\psi|\phi\rangle| = 1$—meaning that they are the same state—then the purity of ω_C is maximal, and indeed $\omega_C = |+\rangle\langle+|$ is a pure state in this case. What this shows is that, as claimed, the entanglement of $|\Psi\rangle$ depends upon the overlap of $|\psi\rangle$ and $|\phi\rangle$, and can therefore be used to *measure* it.

We will now see how to use the above insight in order to derive (and interpret) the SDP for fidelity. Let us look instead at the reduced density operator ω_{AC},

$$\omega_{AC} = \mathrm{tr}_B[|\Psi\rangle\langle\Psi|], \tag{3.44a}$$

$$= \frac{1}{2}(\rho \otimes |0\rangle\langle 0| + \sigma \otimes |1\rangle\langle 1| \tag{3.44b}$$

$$+ e^{-i\theta}\mathrm{tr}_B[|\psi\rangle\langle\phi|] \otimes |0\rangle\langle 1| + e^{i\theta}\mathrm{tr}_B[|\phi\rangle\langle\psi|] \otimes |1\rangle\langle 0|) \tag{3.44c}$$

where we used the fact that $\mathrm{tr}_B[|\psi\rangle\langle\psi|] = \rho$ and $\mathrm{tr}_B[|\phi\rangle\langle\phi|] = \sigma$. Being a density operator, we know that $\omega_{AC} \succcurlyeq 0$. Moreover, it can be seen that the trace of the Hermitian part of the off-diagonal block $e^{-i\theta}\mathrm{tr}_B[|\psi\rangle\langle\phi|]$ is

$$\mathrm{tr}\left[\frac{e^{-i\theta}\mathrm{tr}_B[|\psi\rangle\langle\phi|] + e^{i\theta}\mathrm{tr}_B[|\phi\rangle\langle\psi|]}{2}\right] = |\langle\psi|\phi\rangle|. \tag{3.45}$$

Why is this relevant? Well, recall that, due to Uhlmann's theorem, $|\langle\psi|\phi\rangle|$, when maximised, is precisely $\sqrt{F(\rho, \sigma)}$. We have however just realised that $|\langle\psi|\phi\rangle|$ is also the trace of the Hermitian part of the off-diagonal block of ω_{AC}. This provides us with a way of obtaining an *upper bound* on the fidelity. The off-diagonal block $e^{-i\theta}\mathrm{tr}_B[|\psi\rangle\langle\phi|]$ can be treated as a *variable*, which we denote by $Y + iZ$ (with $Y = Y^\dagger$ and $Z = Z^\dagger$ Hermitian operators) and then consider maximising the trace of Y (the Hermitian part), subject to the constraint that ω_{AC} is a valid density operator, namely

$$\sqrt{F(\rho, \sigma)} \leqslant \text{ maximise } \quad \mathrm{tr}(Y) \tag{3.46a}$$

$$\text{subject to } \quad \begin{pmatrix} \rho & Y + iZ \\ Y - iZ & \sigma \end{pmatrix} \succcurlyeq 0, \tag{3.46b}$$

where we have written ω_{AC} as a block matrix, and have ignored the overall factor of $\frac{1}{2}$, which doesn't affect whether or not it is positive semidefinite. This is an upper bound, since we know that it is possible to take $Y + iZ = e^{-i\theta}\mathrm{tr}_B[|\psi\rangle\langle\phi|]$, with $|\psi\rangle$ and $|\phi\rangle$ the optimal purifications of ρ and σ respectively (which achieve the fidelity through Uhlmann's theorem), and with θ such that $e^{i\theta}\langle\psi|\phi\rangle$ is real. However, since we have relaxed the problem, and are maximising, there is no guarantee that this is the optimal solution, the maximum might be attained at a strictly larger number than the fidelity.

In order to show that this isn't the case, we can now proceed in the other direction, to show that $\mathrm{tr}(Y^*) \leqslant \sqrt{F(\rho, \sigma)}$. Let us assume that (3.46) has been solved, and that optimal variables Y^* and Z^* have been found. This in particular means that we have found a density operator

$$\omega^*_{AC} = \frac{1}{2}(\rho \otimes |0\rangle\langle 0| + \sigma \otimes |1\rangle\langle 1| + (Y^*+iZ^*) \otimes |0\rangle\langle 1| + (Y^*-iZ^*) \otimes |1\rangle\langle 0|). \quad (3.47)$$

Consider then a purification of ω^*_{AC}, which we denote $|\Psi^*\rangle$ (with the purifying system labelled B). From this purification we can define two (normalised) states,

$$|\psi^*\rangle = \sqrt{2}(\mathbb{I}_{AB} \otimes \langle 0|_C)|\Psi^*\rangle, \qquad |\phi^*\rangle = \sqrt{2}(\mathbb{I}_{AB} \otimes \langle 1|_C)|\Psi^*\rangle, \quad (3.48)$$

Crucially, these states can be seen to be purifications of ρ and σ. Indeed, we see that

$$\mathrm{tr}_B[|\psi^*\rangle\langle\psi^*|] = 2\mathrm{tr}_B[(\mathbb{I}_{AB} \otimes \langle 0|_C)|\Psi^*\rangle\langle\Psi^*|(\mathbb{I}_{AB} \otimes |0\rangle_C)], \quad (3.49a)$$

$$= 2[(\mathbb{I}_A \otimes \langle 0|_C)\omega^*_{AC}(\mathbb{I}_A \otimes |0\rangle_C)], \quad (3.49b)$$

$$= \rho, \quad (3.49c)$$

and similarly $\mathrm{tr}_B[|\phi^*\rangle\langle\phi^*|] = \sigma$. (Note also that this confirms that $|\psi^*\rangle$ and $|\phi^*\rangle$ are normalised.) Now, the overlap between $|\psi^*\rangle$ and $|\phi^*\rangle$ is

$$|\langle\psi^*|\phi^*\rangle| = 2\langle\Psi^*|(\mathbb{I}_{AB} \otimes |0\rangle\langle 1|_C)|\Psi^*\rangle, \quad (3.50a)$$

$$= 2|\mathrm{tr}[(\mathbb{I}_A \otimes |0\rangle\langle 1|_C)\omega^*_{AC}]|, \quad (3.50b)$$

$$= |\mathrm{tr}(Y^*) + i\mathrm{tr}(Z^*)|, \quad (3.50c)$$

$$= \sqrt{\mathrm{tr}(Y^*)^2 + \mathrm{tr}(Z^*)^2}. \quad (3.50d)$$

So, from Ulhmann's theorem, since $|\psi^*\rangle$ and $|\phi^*\rangle$ are purifications of ρ and σ, we know that the fidelity $F(\rho, \sigma)$ must be *at least* $|\langle\psi^*|\phi^*\rangle|^2$, and therefore we have the bound

$$\sqrt{F(\rho, \sigma)} \geqslant |\langle\psi^*|\phi^*\rangle| = \sqrt{\mathrm{tr}(Y^*)^2 + \mathrm{tr}(Z^*)^2}. \quad (3.51)$$

However, comparing with (3.46), we have $\mathrm{tr}(Y^*) \geqslant \sqrt{F(\rho, \sigma)}$. The only way that these two bounds can be satisfied is if $\sqrt{F(\rho, \sigma)} = \mathrm{tr}(Y^*)$ and $\mathrm{tr}(Z^*) = 0$.

That is, this shows that in (3.46) the inequality can be replied by an equality, and that in fact this SDP precisely evaluates to the fidelity between ρ and σ.

The derivation is moreover interesting since it sheds light on how to interpret the SDP. We can view the objective function as maximising the entanglement of a state of the form (3.41), which entangles purifications of ρ and σ. Only if the states have a small fidelity (meaning they are close to being orthogonal) can there exist purifying states that then can be superposed and generate a large amount of entanglement. When the states have a large fidelity, meaning they are similar, it is not possible to find purifications of them which generate much entanglement by superposing them.

This can be viewed as a type of *trade-off* relation, between how similar states are, and how much their purifications can be entangled with an auxiliary system, and shows that the fidelity quantifies this trade-off.

IOP Publishing

Paul Skrzypczyk and Daniel Cavalcanti

Chapter 4

Quantum measurements

In this chapter we will see that semidefinite programming is highly relevant and applicable to problems involving optimisation over *quantum measurements*. The reason for this is similar to the reason why semidefinite programs (SDPs) arise so broadly when considering problems involving quantum states—because the elements of a measurement $\mathbb{M} = \{M_a\}$ are positive semidefinite operators, $M_a \succcurlyeq 0$, and the normalisation condition, $\sum_a M_a = \mathbb{1}$, is a linear equality constraint. This means that it is always possible to optimise over of the set of quantum measurements inside an SDP. In this chapter, we will see numerous examples of problems involving measurements where SDP techniques can be applied to solve the problem, and to gain important insights.

In particular, we will begin by considering the analogous problem to state estimation considered in the previous chapter—namely measurement estimation. It will be seen that many of the same ideas carry over in a very natural way to this setting. We will then turn to a second fundamental task of quantum information science, which finds application in numerous settings: the problem of quantum state discrimination. We will see in this chapter how this task can be formulated as a semidefinite program, and that complementary slackness can be used in order to re-derive the well-known optimality conditions which constitute necessary and sufficient conditions satisfied by any measurement which can perform state discrimination optimally.

We then study state discrimination from a second, complementary perspective, considering instead a *fixed* measurement. We will see that semidefinite programming can be used to answer the question of 'how useful' a given measurement is for quantum state discrimination.

4.1 Quantum measurement estimation

Chapter 3 started by considering the problem of quantum state estimation, which was motivated by a situation where an experimentalist observes a set of expectation values when measuring a set of (known) observables on an (unknown) quantum state that they want to estimate. Here, we will begin by considering the symmetric

doi:10.1088/978-0-7503-3343-6ch4 © IOP Publishing Ltd 2023

situation. Namely, consider a situation where an experimentalist uses an unknown measuring device to measure a collection of known quantum states. As previously, the question now is to determine which set of measurements is compatible with the experimental data.

In more detail, let us assume that we have an unknown measurement $\mathbb{M} = \{M_a\}_a$, with outcomes $a = 1, \ldots, o$, which is used to measure a known set of N quantum states $\{\rho_x\}_x$, leading to observed statistics $\{p(a|x)\}_{a,x}$. Finding a measurement compatible with these statistics is given by the following feasibility SDP[1]:

$$\text{find} \quad \mathbb{M} \tag{4.1a}$$

$$\text{subject to} \quad \text{tr}(M_a\rho_x) = p(a|x) \qquad \forall\, a,\, x, \tag{4.1b}$$

$$M_a \succcurlyeq 0 \qquad a = 1, \ldots, o, \tag{4.1c}$$

$$\sum_a M_a = \mathbb{1}. \tag{4.1d}$$

As in the case of (3.1), there are three possibilities that can arise: if the set of states $\{\rho_x\}_x$ is *tomographically complete*—meaning that measurements on such a set of states are sufficient to uniquely determine the measurement performed—then if the statistics were feasible, there will be a *unique solution*; on the other hand, if the states are not tomographically complete, but the statistics are still compatible, then there will be infinitely many solutions; finally, if the statistics cannot be produced by any measurement, then the problem is infeasible.

In the above, we have assumed that the full statistics $p(a|x)$ are given. It is possible consider other possibilities for the data that is given, for example to only be given expectation values. In this case, the problem remains a feasibility SDP. This formulation is studied in exercise 4.1.

We can also consider other variants of the problem, similar to those considered in chapter 3. For example, it is natural that the probabilities $p(a|x)$ will not be known exactly, but will have some inherent uncertainty, due to finite statistics. If we denote the uncertainty in the probability $p(a|x)$ by Δ_{ax}, then the following feasibility SDP searches for a valid measurement consistent with these constraints:

$$\text{find} \quad \mathbb{M} \tag{4.2a}$$

$$\text{subject to} \quad p(a|x) - \Delta_{ax} \leqslant \text{tr}(M_a\rho_x) \leqslant p(a|x) + \Delta_{ax} \qquad \forall\, a,\, x, \tag{4.2b}$$

$$M_a \succcurlyeq 0 \qquad a = 1, \ldots, o, \tag{4.2c}$$

$$\sum_a M_a = \mathbb{1}. \tag{4.2d}$$

[1] Note that in (4.1b) we have written $\forall\, a,\, x$ as shorthand for $a = 1, \ldots, o$; $x = 1, \ldots, N$, to avoid unnecessary clutter. In the rest of this book, whenever a set of conditions holds for more than one variable in a given line of an SDP, we will always use this shorthand, with the range left implicit.

Finally, it can also be useful to work with a standard optimisation SDP, rather than the feasibility SDP (4.1). One way to achieve this, which leads to an SDP that is bounded and feasible—which will therefore have a dual formulation—is to consider minimising a uniform perturbation of the probabilities $p(a|x)$. Namely, the following minimisation SDP

$$\text{minimise} \quad \delta \tag{4.3a}$$

$$\text{subject to} \quad p(a|x) - \delta \leqslant \operatorname{tr}(M_a\rho_x) \leqslant p(a|x) + \delta \qquad \forall a, x, \tag{4.3b}$$

$$M_a \geqslant 0 \qquad a = 1, \ldots, o, \tag{4.3c}$$

$$\sum_a M_a = \mathbb{1}. \tag{4.3d}$$

If the feasibility problem (4.1) was feasible, then $\delta^* = 0$, and both the left-hand and right-hand inequality constraints in (4.3b) become equal, and so enforce that $\operatorname{tr}(M_a\rho_x) = p(a|x)$. On the other hand, if (4.1) was infeasible, then $\delta^* > 0$ will be necessary in order to find a solution. However, it is clear that it will always be possible to find a solution with $\delta^* \leqslant 1$, since at $\delta = 1$, the constraints (4.3b) are necessarily vacuous, and any measurement will be a feasible solution.

In exercise 4.2 the dual formulation of (4.3) is derived. It is shown that this formulation is

$$\text{maximise} \quad \sum_{a,x}(v_{ax} - u_{ax})p(a|x) + \operatorname{tr}(Y) \tag{4.4a}$$

$$\text{subject to} \quad Y + \sum_x(v_{ax} - u_{ax})\rho_x \preccurlyeq 0 \quad a = 1, \ldots, o, \tag{4.4b}$$

$$\sum_{a,x}(u_{ax} + v_{ax}) = 1 \tag{4.4c}$$

$$u_{ax} \geqslant 0, \quad v_{ax} \geqslant 0 \qquad \forall a, x. \tag{4.4d}$$

This dual formulation can be understood geometrically, such that we arrive at a very similar type of situation as encountered in figure 3.2, in the context of state estimation. In particular, let us group all of the probabilities $p(a|x)$ into a single vector $\vec{p}$. That is, we will have a vector whose components are $p(a|x)$, and therefore, it will be natural to label the components by the pair of indices a and x, which will be denoted p_{ax}.[2] In a similar fashion, we can define vectors $\vec{u}$ and $\vec{v}$ with components u_{ax} and v_{ax} respectively. Recalling (1.62), it can be seen that it is advantageous to

[2] If this is confusing, you can imagine that components are labelled by i, and that the ordering of the components runs through all possibility probabilities, i.e. $\vec{p} = (p(1|1), p(2|1), \ldots, p(o|1), p(1|2), \ldots, p(o|N))$. However, it is more convenient, and natural, to simply label the components by the pair, and to leave this ordering as implicitly understood.

combine these vectors into a single vector $\vec{s} = \vec{v} - \vec{u}$, in which case the constraints (4.4c) and (4.4d) can be interpreted as imposing $\|\vec{s}\|_1 \leqslant 1$. This leads us to the following equivalent expression for the dual formulation (4.4),

$$\text{maximise} \quad \vec{s} \cdot \vec{p} + \text{tr}(Y) \tag{4.5a}$$

$$\text{subject to} \quad Y + \sum_x s_{ax}\rho_x \preccurlyeq 0 \qquad a = 1, \ldots, o, \tag{4.5b}$$

$$\|\vec{s}\|_1 \leqslant 1. \tag{4.5c}$$

This is similar in form to (3.18), and can be understood analogously. The data $p(a|x)$ specifies a point $\vec{p}$ in the space of measurement results. Within this space, given the set of states $\{\rho_x\}_x$, there is a (convex) subset of *feasible data*. The dual formulation can be interpreted as finding a *separating hyperplane*, such that if the proposed data $p(a|x)$ was infeasible, it will lie on one side of the plane, while the entire feasible set of data lies on the other side. This is very similar and completely analogous to figure 3.2 from the previous chapter.

Exercises

4.1 In this exercise we will consider the variant of measurement estimation where we are instead given as data the expectation values m_x, obtained when the measurement $\mathbb{M}$ is performed on state ρ_x.
- (a) Write down the quantum mechanical expression for the expectation value when performing a measurement $\mathbb{M} = \{M_a\}_a$ on a state ρ_x.
- (b) Write down a feasibility SDP that will determine whether a $\mathbb{M}$ exists, that is able to reproduce the expectation value data $\{m_x\}_x$.

4.2 In this exercise we will derive the dual SDP (4.4).
- (a) Write down the Lagrangian associated to (4.3), introducing dual variables u_{ax} and v_{ax} for $a = 1, \ldots, o$ and $x = 1, \ldots, N$, associated to the right-hand and left-hand constraints in (4.3b), X_a associated to (4.3c) and Y associated to (4.3d)
- (b) Use the Lagrangian to arrive at a dual formulation of (4.3), by identifying the required inequality and equality constraints that need to be satisfied in order to make the Lagrangian simultaneously a lower bound on the primal objective function and independent of the primal variables.
- (c) Show that in the dual SDP derived in (b) X_a are slack variables, and can therefore be eliminated. Thus, show that (4.4) is the dual SDP formulation of the measurement estimation problem.

4.2 Quantum state discrimination I

We now turn our attention to quantum state discrimination, which is a fundamental task of quantum information science, which finds application in numerous settings. The techniques of semidefinite programming can be widely applied to this task, as we will see in this section and the next.

In the setting of quantum state discrimination, we have a source of quantum particles which outputs probabilistically one of a number of quantum states. The goal is to correctly identify which state was emitted. More formally, the source produces the quantum state ρ_i, for $i = 1, \ldots, m$ with probability $q(i)$. The source is thus characterised by an *ensemble* $\mathcal{E} = \{q(i), \rho_i\}_i$. We will also define the associated probability vector $\vec{q}$ whose elements are $q_i = q(i)$.

We can view quantum state discrimination as the *encoding* of classical data into quantum states, which then needs to be recovered or *decoded*. The classical information is the random variable i, distributed according to $q(i)$. In order to decode the information stored in the quantum state, it is necessary to perform a measurement, and based upon the result of this measurement to make a guess of the value of i. This will only be a guess, since in general it will not be possible to perfectly identify which state ρ_i was sent, and therefore to identify the value of i. In quantum state discrimination the overarching goal is to guess as well as possible the state, and hence the value of i.

There are a number of different figures of merit that one can consider optimising in this task. Here we will focus on the two most widely studied, *minimum-error* quantum state discrimination, and *unambiguous* quantum state discrimination. In the former, we seek to minimise the *average error* made, which is equivalent to maximising the average success probability of correctly guessing; in the latter we seek to maximise the probability of *being certain* that the state has been correctly identified.

4.2.1 Minimum-error quantum state discrimination

The first variant of quantum state discrimination we will consider is when the goal is to minimise the average error of incorrectly identifying the state. Due to symmetry, this can alternatively be phrased as maximising the average probability of correctly identifying the state. Upon receiving a state, we assume that a measurement $\mathbb{M} = \{M_a\}_a$ is performed. If the outcome of the measurement is a, then we will guess that this was the state received. The probability to obtain the outcome a when the state is ρ_i is

$$p(a|i) = \mathrm{tr}(M_a\rho_i). \tag{4.6}$$

The figure of merit we consider is the average probability of correctly guessing the state,

$$P_{\mathrm{guess}} = p(a = i) \tag{4.7a}$$

$$= \sum_i q(i)p(a = i|i) \tag{4.7b}$$

$$= \sum_i q(i)\mathrm{tr}(M_i\rho_i). \tag{4.7c}$$

This is the average probability of guessing correctly using the measurement $\mathbb{M}$, and the guessing strategy that $a = i$. In order to find the best possible average guessing probability, we need to optimise over all choices of measurement, namely

$$P^*_{\text{guess}} = \max_{\mathbb{M}} \sum_i q(i)\text{tr}(M_i\rho_i). \tag{4.8}$$

In exercise 4.3 it is shown that this is indeed the best possible average guessing probability, i.e. that you do not need to optimise over more general guessing strategies, involving, for example, not merely using the measurement result to directly guess which state was sent, $i = a$. The final step is to realise that the optimisation over measurements can be written as an SDP. In particular, using the conditions that must be satisfied by any valid measurement, we arrive at the following explicit SDP formulation of (4.8):

$$P^*_{\text{guess}} = \text{maximise} \quad \sum_i q(i)\text{tr}(M_i\rho_i) \tag{4.9a}$$

$$\text{subject to} \quad M_a \succcurlyeq 0 \qquad a = 1, \dots, m, \tag{4.9b}$$

$$\sum_a M_a = \mathbb{1}. \tag{4.9c}$$

As a first application, we will see below that in the simplest possible quantum state discrimination task—involving two equally likely states—then we can use insights from chapter 2 to solve the SDP analytically, and in so doing, arrive at an operational interpretation for the trace distance (3.2).

4.2.2 Binary state discrimination

In the case of binary state discrimination, where either the state ρ_1 or ρ_2 is given, with equal probability $q(1) = q(2) = 1/2$, we need to optimise over two outcome measurements $\mathbb{M} = \{M_1, M_2\}$. Due to the normalisation constraint $M_1 + M_2 = \mathbb{1}$, we see that only a single measurement operator is actually optimised over. We could solve for $M_2 = \mathbb{1} - M_1$, however this would *break the symmetry* between the two measurement operators somewhat. A more symmetrical way to reduce down to a single operator is to introduce an auxiliary operator Z such that

$$M_1 = \frac{\mathbb{1} + Z}{2}, \qquad M_2 = \frac{\mathbb{1} - Z}{2}. \tag{4.10}$$

By construction, we have that $M_1 + M_2 = \mathbb{1}$. In order for each element to remain positive semidefinite, we see that Z must satisfy

$$-\mathbb{1} \preccurlyeq Z \preccurlyeq \mathbb{1}, \tag{4.11}$$

with the left-hand inequality ensuring that $M_1 \succcurlyeq 0$, and the right-hand ensuring that $M_2 \succcurlyeq 0$. For binary quantum state discrimination, in terms of this new variable, we therefore find that (4.9) becomes

$$P^*_{\text{guess}} = \text{maximise} \quad \frac{1}{2} + \frac{1}{4}\text{tr}[Z(\rho_1 - \rho_2)] \tag{4.12a}$$

$$\text{subject to} \quad -\mathbb{1} \preccurlyeq Z \preccurlyeq \mathbb{1}. \tag{4.12b}$$

At this stage, we can now cast our mind back to exercise 2.6, and the SDP for the trace norm given in (2.11). By inspection, we see that by considering $A = \rho_1 - \rho_2$, the only difference between the two SDPs would be in the objective function, which has been shifted and re-scaled from $\text{tr}(ZA)$ to $\frac{1}{2} + \frac{1}{4}\text{tr}(ZA)$, and therefore we have

$$P^*_{\text{guess}} = \frac{1}{2} + \frac{1}{4}\|\rho_1 - \rho_2\|_1, \tag{4.13a}$$

$$= \frac{1}{2}\big[1 + T(\rho_1, \rho_2)\big] \tag{4.13b}$$

where in the second line we have furthermore used the definition of the trace distance, namely $T(\rho_1, \rho_2) = \frac{1}{2}\|\rho_1 - \rho_2\|_1$ as given in (3.2). We have thus recovered the well-known result—a special instance of the Helstrom bound—the trace distance quantifies how well two states can be distinguished from each other in the setting of minimum-error binary quantum state discrimination. This result is important as it provides a clear *operational interpretation* of the trace distance between two states.

Exercises

4.3 In this exercise we will show that when optimising over measurement strategies in minimum-error quantum state discrimination, without loss of generality we can assume that a fixed measurement is performed, and take the outcome of the measurement to be the guess of which state was received. Consider the following general class of strategies: (i) generate a random variable λ according to a probability distribution $p(\lambda)$; (ii) conditional on λ, when the state is received, apply a unitary transformation U_λ to the state; (iii) measure the state using an arbitrary measurement $\mathbb{M} = \{M_a\}$ and obtain an outcome a; (iv) conditional on both a and λ, probabilistically make a guess for the state according to a probability distribution $p(i|a, \lambda)$.

(a) Write down P_{guess} for the above strategy.

(b) Show that your answer to part (a) can be expressed in a form identical to (4.7c) using a measurement $\mathbb{M}' = \{M'_i\}$, with elements

$$M_i' = \sum_{a,\lambda} p(\lambda)p(i|a, \lambda)U_\lambda^\dagger M_a U_\lambda. \tag{4.14}$$

(c) Verify that $\mathbb{M}'$ from part (b) is a valid measurement.

(d) Explain why the above shows that there is no loss of generality in restricting to the basic subset of strategies considered in the main text.

4.4 Show that for *non-symmetric* binary quantum state discrimination, with arbitrary $q(1)$ and $q(2)$, the average success probability is

$$p^*_{\text{guess}} = \frac{1}{2} + \frac{1}{2}\|q(1)\rho_1 - q(2)\rho_2\|_1$$

4.5 Use exercise 2.7 to show that an optimal measurement for binary quantum state discrimination is

$$M_1^* = \Pi^{(+)}, \qquad M_2^* = \Pi^{(-)}, \tag{4.15}$$

where $\Pi^{(+)}$ and $\Pi^{(-)}$ are the projectors onto the positive and negative parts of $\rho_1 - \rho_2$, respectively, as defined in (2.14). *This provides a complete understanding of the problem of binary quantum state discrimination, giving both the optimal success probability, and the measurement that should be performed in order to achieve it.*

4.2.3 Optimality conditions

Going beyond the simple case of binary quantum state discrimination, it is in general a difficult problem to find the optimal success probability and measurements for minimum-error quantum state discrimination. There are however a set of necessary and sufficient conditions, which if satisfied, guarantee that a set of measurements is indeed optimal. These conditions are

$$M_i^*\big(q(i)\rho_i - q(j)\rho_j\big)M_j^* = 0 \qquad i, j = 1, \dots, m, \tag{4.16a}$$

$$\sum_j q(j)\rho_j M_j^* - q(i)\rho_i \succcurlyeq 0 \qquad i = 1, \dots, m. \tag{4.16b}$$

We will now see that the concept of complementary slackness—introduced in section 2.4—can be used to derive these conditions. Our starting point is the Lagrangian associated to (4.9), which we see is

$$\mathcal{L} = \sum_i q(i)\mathrm{tr}(M_i\rho_i) + \sum_i \mathrm{tr}(Z_i M_i) + \mathrm{tr}\left[W\left(\mathbb{1} - \sum_i M_i\right)\right], \tag{4.17a}$$

$$= \sum_i \mathrm{tr}\big[M_i\big(q(i)\rho_i + Z_i - W\big)\big] + \mathrm{tr}(W), \tag{4.17b}$$

where we have introduced dual variables $Z_i \succcurlyeq 0$ for $i = 1, \dots, m$ and W associated to the positivity constraints (4.9b) and normalisation constraint (4.9c) respectively. The dual SDP formulation for minimum-error quantum state discrimination is therefore

$$P^*_{\text{guess}} = \text{minimise} \quad \mathrm{tr}(W) \tag{4.18a}$$

$$\text{subject to} \quad q(i)\rho_i + Z_i - W = 0 \qquad i = 1, \dots, m, \tag{4.18b}$$

$$Z_i \succcurlyeq 0 \qquad i = 1, \dots, m. \tag{4.18c}$$

We have chosen not to eliminate the slack variables Z_i for the time being. The reason for doing so will become clear below, but we emphasise that when using complementary slackness, it is important to keep all dual variables initially, even the slack variables.

First of all, in order to use complementary slackness it necessary to check that strong duality holds. Here, it can be seen directly that it does indeed hold. In particular, note first that for all values of a, $M_a = \mathbb{1}/m \succ 0$ is a strictly feasible solution for the primal SDP (4.9). Second, for all values of i, $Z_i = \gamma\mathbb{1} \succ 0$ is a strictly feasible dual variables for the dual SDP (4.18). Since both problems are thus strictly feasible, strong duality holds. This confirms in turn that P^*_{guess} is finite, as we know to be true on physical grounds.

Since strong duality holds, we can now appeal to complementary slackness, in order to see what can be learnt about the optimal variables. From (4.17), we see that the second term must vanish in order that $\mathcal{L} = p^*_{\text{guess}}$, which implies that

$$Z_i^* M_i^* = 0 \quad i = 1, \ldots, m, \tag{4.19}$$

since $M_i^* \succcurlyeq 0$ and $Z_i^* \succcurlyeq 0$, as explained in section 2.4.

Now, from the dual constraint (4.18b), we see that for all i, $Z_i^* = W^* - q(i)\rho_i$. Inserting this into the complementary slackness condition, we see that

$$0 = Z_i^* M_i^* = (W^* - q(i)\rho_i) M_i^* \tag{4.20a}$$

$$\Rightarrow W^* M_i^* = q(i)\rho_i M_i^* \qquad i = 1, \ldots, m. \tag{4.20b}$$

If we now sum both sides of (4.20b) over i, given that W^* is independent of i, and that $\sum_i M_i^* = \mathbb{1}$, we can solve for W^* in terms of the optimal dual measurement operators M_i^*, namely

$$W^* = \sum_i q(i)\rho_i M_i^*. \tag{4.21}$$

This is already an important relationship between the optimal primal and dual variables. Using the fact again that $Z_i^* = W^* - q(i)\rho_i$, combined with the above, and the constraint $Z_i^* \succcurlyeq 0$, we see that

$$Z_i^* = \sum_i q(i)\rho_i M_i^* - q(i)\rho_i \succcurlyeq 0, \tag{4.22}$$

which is precisely the second optimality condition (4.16b).

To obtain the first optimality condition we start by returning to (4.20b), and consider multiplying both sides of the equation from the left by an optimal measurement operator M_j^*, which shows that

$$M_j^* W^* M_i^* = M_j^* q(i)\rho_i M_i^* \quad i, j = 1, \ldots, m. \tag{4.23}$$

We can now consider taking the Hermitian conjugate of this equation, keeping in mind that all of the operators involved are Hermitian. Interchanging the indices $i \leftrightarrow j$, we arrive at

$$M_j^* W^* M_i^* = M_j^* q(j)\rho_j M_i^* \quad i, j = 1, \ldots, m, \tag{4.24}$$

which is almost identical to before, except now on the left-hand side we have $q(j)\rho_j$ instead of $q(i)\rho_i$. Crucially, since the left-hand sides of (4.23) and (4.24) are equal, this allows us to eliminate the slack variable W^*, and obtain

$$M_i^*\big(q(i)\rho_i - q(j)\rho_j\big)M_j^* = 0 \quad i, j = 1, \ldots, m, \tag{4.25}$$

which are precisely the first set of optimality conditions (4.16).

Thus, we have shown that the optimality conditions (4.16) are implications of the complementary slackness conditions associated to the primal and dual SDP formulations of minimum-error quantum state discrimination. As a small note, the complementary slackness conditions themselves are both necessary and sufficient. Since these were used to derive implications, we recover the fact that (4.16) are *necessary* conditions that must be satisfied. What is not shown yet, is that they remain *sufficient* conditions, equivalent to the complementary slackness conditions. This 'converse' argument is however straightforward, and left for exercise 4.8 below.

Exercises

4.6 Show that the optimal measurements for binary quantum state discrimination found in exercise 4.5 satisfy the optimality conditions (4.16) *as they must*.

4.7 Calculate the optimal dual variables W^* and Z_i^* for $i = 1, 2$, for binary quantum state discrimination, and show that the complementary slackness conditions (4.19) are satisfied for these in combination with the optimal measurements found in exercise 4.5.

4.8 In this exercise we will show that the optimality conditions (4.16) are equivalent to the complementary slackness conditions. In particular, it needs to be shown that starting from the conditions (4.16), a feasible set of optimal dual variables can be defined, that satisfy complementary slackness (4.19). Note that we do not need to worry about the primal variables being feasible, since by assumption a valid measurement has been found for the conditions to make sense. We will guess, based upon (4.21) and (4.22), that the following set of dual variables are optimal:

$$W^* = \sum_i q(i)\rho_i M_i^*, \qquad Z_i^* = \sum_j q(j)\rho_j M_j^* - q(i)\rho_i \tag{4.26}$$

(a) Show that, due to the second set of optimality constraints (4.16b), these dual variables are dual feasible, i.e. satisfy (4.18b) and (4.18c).

(b) Show that, due to the first set of optimality constraints (4.16), the complementary slackness conditions (4.19), $Z_i^* M_i^* = 0$, are satisfied. *Hint: It will be useful to use the fact that* $Z_i^{*\dagger} = Z_i^*$ *in order to obtain an alternative expression for* Z_i^* *as a first step.*

Together, parts (a) and (b) imply that whenever the optimality conditions (4.16) *are satisfied, then so too are the complementary slackness conditions. Since satisfying complementary slackness is sufficient for being optimal, this shows that the optimality conditions are both necessary and sufficient.*

4.2.4 Unambiguous quantum state discrimination

We now consider a second variant of quantum state discrimination, known as *unambiguous quantum state discrimination*. In the previous setting, of minimum-error quantum state discrimination, all that was required was that *on average* the state was identified as well as possible. This means, in particular, that *errors* are made: the guess of the state is not always equal to the state received. In some circumstances, such errors can be problematic or highly undesirable, and this motivates unambiguous state discrimination.

In the current setting, we consider allowing a new freedom in state discrimination —the possibility of *not making a guess of the state*, that is, of declaring failure in identifying the state. Now, however, we demand that whenever a guess is made, we should be *certain* that the state has been correctly identified. The figure of merit in unambiguous state discrimination is therefore to maximise the probability of being certain about the state, which is equivalent to minimising the probability of being uncertain about the state (i.e. declaring failure).

In order to model this situation, when considering a quantum state discrimination task with m states, a measurement $\mathbb{M} = \{M_a\}$ with $m + 1$ outcomes is used. The first m outcomes are the guesses, $a = 1, \dots, m$, while the additional outcome $a = \emptyset$ is the 'failure' outcome. That is, the result $a \neq \emptyset$ should be returned whenever it is certain that the corresponding state was sent—i.e. when the state has been *unambiguously* identified—and should return the failure outcome $a = \emptyset$ in any other case, signifying the failure to identify the state.

How can we guarantee that when an outcome $a \neq \emptyset$ is seen that we are certain of the state? In order to do so, it must be the case that the probability of seeing any outcome other than i on the state ρ_i must be strictly zero. That is, the measurement used must be restricted to satisfy

$$p(a|i) = \mathrm{tr}(M_a\rho_i) = 0 \quad i, a = 1, \dots, m; \; a \neq i. \tag{4.27}$$

Note that this condition is not imposed for $M_\emptyset$, the measurement element associated with failure. For reasons that will become clear below, it will be useful to consider instead of the state ρ_i, the product, or unnormalised state, $q(i)\rho_i$ namely to impose

$$p(a, i) = q(i)\mathrm{tr}(M_a\rho_i) = 0 \quad i, a = 1, \dots, m; \; a \neq i. \tag{4.28}$$

One way to see why this is in fact a more natural constraint to impose is that it means that if a state would occur with zero probability, $q(i) = 0$, for some i, then by imposing (4.28) no constraint would be imposed, while according to (4.27) a constraint would still be imposed, which is rather unnatural. We will also see below that if (4.28) is used, it will lead to a simpler dual SDP with a more natural structure (compared to using (4.27)), which will prove useful.

Since (4.28) constitute a set of linear constraints that must be satisfied by the measurement, we can impose them as constraints inside an SDP—restricting the class of measurements considered—and optimise to find the best such measurement for unambiguous state discrimination. We therefore arrive at the following SDP which provides the largest average probability of unambiguously identifying a state

$$P^*_{\text{guess}} = \text{maximise} \quad \sum_{a=1}^{m} q(i)\text{tr}(M_i\rho_i) \tag{4.29a}$$

$$\text{subject to} \quad q(i)\text{tr}(M_a\rho_i) = 0 \qquad i, a = 1, \ldots, m;\ a \neq i. \tag{4.29b}$$

$$M_a \succcurlyeq 0 \qquad a = 1, \ldots, m, \varnothing \tag{4.29c}$$

$$\sum_a M_a = \mathbb{1}. \tag{4.29d}$$

Unambiguous state discrimination is a harder task than minimum-error state discrimination, since the requirements on the measurement are more demanding. It can be seen that in certain circumstances unambiguous state discrimination is impossible, meaning that the optimal measurement for a given ensemble of states $\mathcal{E} = \{q(i), \rho_i\}$ has $M_a^* = 0$ for $a = 1, \ldots, m$ and $M_\varnothing^* = \mathbb{1}$, leading to $P^*_{\text{guess}} = 0$.

We can actually understand when this happens. There are two separate cases to consider. First, since $\rho_i \succcurlyeq 0$, the only way that $\text{tr}(M_a\rho_i) = 0$ is if $M_a\rho_i = 0$. Therefore, if a ρ_i is full rank, then it follows that no other state can be unambiguously identified, $M_a = 0$, for $a \neq i$. Unambiguous state discrimination is therefore impossible for a set of two or more full-rank states, and is uninteresting already when one state is full rank. We therefore restrict our attention to the case where none of the states are full rank.

Interestingly, even if none of the states are full rank it can still happen that unambiguous state discrimination is impossible. In the 'Advanced topics' section 4.6.1 we investigate this further, and show how it can lead to insight into the dual SDP for unambiguous state discrimination.

4.3 Quantum state discrimination II

In the previous section we viewed quantum state discrimination from the perspective where the ensemble of states to be discriminated was the *data* of the problem, and considered optimising over the set of quantum measurements, in order to identify the state with the largest probability of success (or, said differently, with the smallest probability of error).

Quantum state discrimination is also interesting from an alternative perspective, where we view the measurement as *fixed*, i.e. as the *input data*, and allow ourselves to optimise over different state discrimination tasks, each specified by an ensemble $\mathcal{E}$. In this section, we will consider quantum state discrimination from this alternative perspective.

The motivation for considering this comes from understanding the question of *how informative a measurement is*—about its ability to provide classical information about the quantum state. Intuitively, one way to see that a given measurement is highly informative is if it can perform very well in at least one quantum state discrimination task. On the other hand, if a measurement is unable to perform well in any quantum state discrimination task, then this is naturally an uninformative measurement.

We will see below that this intuition can be formalised, and that the duality of semidefinite programming shows that quantum state discrimination provides an operational characterisation of a quantifier of measurement informativeness, known as the *generalised robustness of measurement informativeness*. This is a particular example of a more general correspondence that exists, and it is thus useful to study this example in detail to learn how SDPs can be applied in such settings.

Let us focus on the task of minimum-error quantum state discrimination, as considered in section 4.2.1. Assuming the same guessing strategy as previously—that if the outcome a is observed, a guess is made that the state received was ρ_a—the probability of correctly guessing the state, for a fixed measurement $\mathbb{M} = \{M_a\}$ with o outcomes, $a = 1, \dots, o$, and for a fixed ensemble $\mathcal{E} = \{q(i), \rho_i\}$, with the same number of states o, is

$$P_{\text{guess}} = \sum_i q(i)\text{tr}(M_i\rho_i). \tag{4.30}$$

Inspired by the analysis from before, we may be tempted to optimise this average success probability over all ensembles in order to define a figure of merit for how informative a measurement is, namely

$$P^*_{\text{guess}} = \text{maximise} \quad \sum_i q(i)\text{tr}(M_i\rho_i), \tag{4.31}$$

that is, to merely interchange the role played by the measurement (now considered the fixed input data), and the ensemble (now considered the optimisation variable). However, as shown in exercise 4.11, P^*_{guess} is not a good figure of merit, since it is maximised by an ensemble *with only a single state occurring with unit probability*. What is problematic about (4.31) is that it doesn't capture the fact that as the ensemble $\mathcal{E}$ is varied, the intrinsic difficultly in identifying the state also varies. Consider for example a trivial situation in which we have no ability to perform a measurement at all. How well can we do in a quantum state discrimination task? The optimal strategy is to *guess the most likely state all of the time*. If we do this, our probability of correctly guessing the state is

$$P^{(\text{no meas})}_{\text{guess}} = \max_i q(i). \tag{4.32}$$

Therefore, we can always do very well in a state discrimination game that is very asymmetric, such that one state occurs very frequently. In the limiting case, where only a single state is sent, with unit probability, the game can be won (trivially) with certainty, irrespective of the measurement we use. Thus, the optimisation problem (4.31) can be maximised by picking games of this type.

To overcome the above problem, we instead consider the *advantage* that is offered by a measurement in a particular game, compared to what can be achieved with no measurement. Said differently, we can consider normalising P_{guess} by $P^{(\text{no meas})}_{\text{guess}}$, namely

$$\text{maximise} \quad \frac{\sum_i q(i)\mathrm{tr}(M_i\rho_i)}{\max_j q(j)} \tag{4.33a}$$

$$\text{subject to} \quad q(i) \geqslant 0 \quad i = 1, \ldots, o, \quad \sum_i q(i) = 1, \tag{4.33b}$$

$$\rho_i \geqslant 0, \qquad \mathrm{tr}(\rho_i) = 1 \qquad i = 1, \ldots, o. \tag{4.33c}$$

This problem has a non-linear objective function, and as such is not directly an SDP. However, as we will now see, an SDP can nevertheless be found which has exactly the same optimal value as (4.33), which can be understood as being equivalent to it. In particular, consider the following SDP

$$\text{maximise} \quad \sum_i \mathrm{tr}(M_i\omega_i) \tag{4.34a}$$

$$\text{subject to} \quad \omega_i \geqslant 0, \qquad \mathrm{tr}(\omega_i) \leqslant 1 \qquad i = 1, \ldots, o. \tag{4.34b}$$

A priori this SDP is unrelated to the previous optimisation problem (4.33). However, to see why they are related, let us first consider that we have found a set of optimal variables $\mathcal{E}^* = \{q^*(i), \rho_i^*\}$ for the original optimisation problem, and let us consider defining a set of potential variables for the SDP (4.34) via

$$\omega_i = \frac{q^*(i)\rho_i^*}{\max_j q^*(j)}. \tag{4.35}$$

In order to be feasible for the original problem (4.33), $q^*(i) \geqslant 0$ and $\rho_i^* \geqslant 0$, and so we see that $\omega_i \geqslant 0$, as required by the first constraint in (4.34b). Moreover, we also see that

$$\mathrm{tr}(\omega_i) = \frac{q^*(i)}{\max_j q^*(j)} \leqslant 1, \tag{4.36}$$

and so the second constraint in (4.34b) is also satisfied. Thus, ω_i is a feasible variable for the SDP (4.34) for all $i = 1, \ldots, o$. Denoting the optimal value of the original problem (4.33) by α^* and the optimal value of the SDP (4.34) by $\tilde{\alpha}^*$, we therefore see that the latter places an upper bound on the original problem, since

$$\alpha^* = \frac{\sum_i q^*(i)\mathrm{tr}(M_i\rho_i^*)}{\max_j q^*(j)} = \sum_i \mathrm{tr}(M_i\omega_i) \leqslant \tilde{\alpha}^*, \tag{4.37}$$

where the first equality is by definition of being optimal for the original (4.33), the second equality follows from the definition of ω_i in (4.35), and the inequality follows since we have no guarantee that this particular choice of ω_i is optimal for the SDP (4.34), and so the value of the objective function only lower bounds the optimal value $\tilde{\alpha}^*$.

We now consider going in the other direction, and assume that an optimal set of variables $\{\omega_i^*\}$ for the SDP (4.34) has been found, and define a set of potential variables for the original problem (4.33) via[3]

$$q(i) = \frac{\mathrm{tr}(\omega_i^*)}{\sum_j \mathrm{tr}(\omega_j^*)}, \qquad \rho_i = \frac{\omega_i^*}{\mathrm{tr}(\omega_i^*)}. \tag{4.38}$$

Since $\omega_i \geqslant 0$ in order to be feasible for the SDP (4.34), it follows that $q(i) \geqslant 0$, and by construction $\sum_i q(i) = 1$, so this is a valid set of probabilities, and is feasible for the original problem (4.33). Similarly, $\rho_i \geqslant 0$, and by construction $\mathrm{tr}(\rho_i) = 1$, so this is also a valid set of states, that is feasible for the original problem (4.33). Altogether, this defines a feasible ensemble $\mathcal{E}$. The advantage that $\mathbb{M}$ provides for the associated discrimination game is

$$\frac{\sum_i q(i)(M_i\rho_i)}{\max_j q(j)} = \frac{\sum_i \mathrm{tr}(M_i\omega_i^*)}{\max_j \mathrm{tr}(\omega_j^*)} \geqslant \sum_i \mathrm{tr}(M_i\omega_i^*) = \tilde{\alpha}^*, \tag{4.39}$$

where the first equality is obtained by substituting (4.38) and simplifying, the middle inequality follows since $\max_j \mathrm{tr}(\omega_j^*) \leqslant 1$ in order to be feasible for the SDP (4.34), and the second equality is by definition of the optimal value of the SDP. Since this ensemble need not be the optimal ensemble for the original problem (4.33), we see that $\alpha^* \geqslant \tilde{\alpha}^*$. Crucially, since it was previously also shown in (4.37) that $\alpha^* \leqslant \tilde{\alpha}^*$, we have inequalities in both directions, and hence we arrive at the important conclusion that

$$\alpha^* = \tilde{\alpha}^*, \tag{4.40}$$

that is, the two problems have the same optimal value. This is very important, as it shows that the advantage that is offered by a measurement in quantum state discrimination, which is naturally a *non-linear* quantity, can in fact be solved by SDP, once it is realised that there is an *associated* SDP. Moreover, when this SDP is solved, not only is the optimal value of the original problem recovered, but via (4.38) so too are the associated optimal variables. We therefore obtain all of the information about the solution.

At this stage, you might be wondering how we arrived at the realisation that such an equivalent SDP existed? It appears to have sprung out of thin air. Will this hold more generally, or was it a genuine piece of magic? Although in general such an associated SDP may not in general exist, or it may be difficult to find even if it does, there are three key steps that were taken here—in the background—that led to this realisation. It will be instructive to go over these three steps, which we do in the next section.

Before doing so, however, we will conclude with the realisation that the SDP (4.34) can be solved rather readily, allowing us to obtain an analytical expression for the advantage provided by a given measurement. We can start by noticing that the SDP *decouples*. That is, each of the variables ω_i, for fixed *i*, is independent of the others, satisfying its own pair of constraints. At the same time, the objective function

[3] Note that if $\omega_i = 0$ for some *i*, then we set $\rho_i = 0$, i.e. we do not encounter the problem of dividing by zero.

is also a linear combination of objective functions for each variable. The optimal value is therefore the sum of the optimal values α_i^* for the problems

$$\text{maximise} \quad \mathrm{tr}(M_i\omega_i) \tag{4.41a}$$

$$\text{subject to} \quad \omega_i \succcurlyeq 0, \qquad \mathrm{tr}(\omega_i) \leqslant 1 \qquad i = 1, \ldots, o. \tag{4.41b}$$

Comparing to (2.1), it is evident that this is *almost* exactly the same SDP seen previously for evaluating the maximum eigenvalue of a Hermitian operator. The only difference is that previously we had the equality constraint $\mathrm{tr}(\omega_i) = 1$, rather than the inequality constraint $\mathrm{tr}(\omega_i) \leqslant 1$. It can however be shown that *without loss of generality* the constraint $\mathrm{tr}(\omega_i) \leqslant 1$ will always be saturated by an optimal variable. Hence we can replace it by the equality constraint. This is shown in exercise 4.10.

This shows that (4.41) evaluates to the maximum eigenvalue of M_i. However, since $M_i \succcurlyeq 0$, we furthermore know that the maximum eigenvalue coincides with the operator norm, $\lambda_{\max}(M_i) = \|M_i\|_\infty$. Hence, we can see that the advantage provided by a measurement in quantum state discrimination is

$$\max_{\mathcal{E}=\{q(i),\rho_i\}} \frac{\sum_i q(i)\mathrm{tr}(M_i\rho_i)}{\max_j q(j)} = \sum_a \|M_a\|_\infty. \tag{4.42}$$

This is a non-trivial function of the measurement $\mathbb{M}$. In exercise 4.9, we show that the most and least informative measurements can be identified using this form, and that the former correspond to measurements with rank-1 elements, while the latter have elements proportional to the identity operator.

Exercises

4.9 In this exercise we will use the form (4.42) to identify the most and least informative measurements.

(a) Consider a measurement $\mathbb{M}$ acting on d-dimensional quantum states with $o \geqslant d$ outcomes. Denote the spectral decomposition of each measurement operator by

$$M_a = \sum_i \lambda_i^{(a)} \Pi_i^{(a)}, \tag{4.43}$$

where $\lambda_i^{(a)} \geqslant 0$, we order the eigenvalues in decreasing order, and where $\Pi_i^{(a)}$ are rank-1 projectors. Show that the advantage offered by a measurement $\mathbb{M}$ is upper bounded, $\alpha^* \leqslant d$. *Hint: You may want to consider taking the trace of the normalisation condition* $\sum_a M_a = \mathbb{1}$.

(b) Consider measurements $\mathbb{M}$ acting on d-dimensional quantum states with $o \geqslant d$ outcomes, for which its measurement elements are of the form $M_a = \gamma_a \Pi_a$, where $\gamma_a > 0$ is a positive constant and Π_a is a rank-1 projector. Show that in this case $\alpha^* = d$.

(c) Conversely, using your analysis from part (a) or otherwise, show that if a measurement $\mathbb{M}$ has even a single measurement operator with two or more non-zero eigenvalues, then $\alpha^* < d$.

(d) Show that $\|M_a\|_\infty$ is minimised when all of the eigenvalues of M_a are equal to each other. Use this to show that the least informative measurements, which provide no advantage in any state discrimination task, have measurement operators of the form $M_a = q(a)\mathbb{1}$, where $q(a)$ is a probability distribution.

4.10 In this exercise we will show that *without loss of generality* $\text{tr}(\omega_i^*) = 1$ in (4.41).

(a) Assume that we have found $\omega_i^* \succcurlyeq 0$ such that $\text{tr}(\omega_i^*) = \delta < 1$ that we *believe* to be optimal. Consider now adding onto ω_i^* another operator $\zeta_i \succcurlyeq 0$, such that $\text{tr}(\omega_i^* + \zeta_i) = 1$, i.e. *any* positive semidefinite operator such that $\text{tr}(\zeta_i) = 1 - \delta > 0$. Show that the new variable $\omega_i' = \omega_i^* + \zeta_i$ is feasible for (4.41) and saturates the second constraint.

(b) Show that the new variable achieves the value

$$\alpha = \text{tr}(M_i\omega_i^*) + \text{tr}(M_i\zeta_i) \geqslant \alpha^*. \tag{4.44}$$

This shows that ω_i' is never a worse variable than ω_i^*; it can only increase the value of the objective function, not decrease it.

(c) Consider now any ζ_i which is not orthogonal to M_i, i.e. such that $\text{tr}(M_i\zeta_i) > 0$. Show that this *must* increase the value of the objective function.

(d) Explain why this is in contradiction with the assumption that ω_i^* was optimal and hence why the optimal variables *must* saturate the inequality $\text{tr}(\omega_i) \leqslant 1$.

4.3.1 Transforming the non-linear problem into a linear one

In this section we will now describe the main ideas that allowed us to arrive at the SDP (4.34). We will start by analysing the objective function of the original problem, (4.33) which is non-linear in two distinct ways. First of all it has a *fractional form*, involving the division of one function by another. Second, from the perspective of this fractional form, both the numerator and denominator are non-linear functions themselves. In order to arrive at a problem with a linear objective function—as is needed for SDPs—we therefore need to understand how to overcome these different contributions to the non-linearity of the objective function.

We will start by realising that the numerator can be made linear by *combining* the variables together, and defining

$$\sigma_i = q(i)\rho_i. \tag{4.45}$$

This can be viewed as a *collection of sub-normalised states*, just as we previously saw when studying unambiguous quantum state discrimination in section 4.2.4. This may appear as a trick, but in fact it is deeper than that. The reason why is because we can map back-and-forth, in a one-to-one fashion, between ensembles $\mathcal{E} = \{q(i), \rho_i\}$ and sets of sub-normalised states $\tilde{\mathcal{E}}=\{\sigma_i\}$, via $q(i) = \text{tr}(\sigma_i)$, $\rho_i = \sigma_i/q(i)$, as shown in exercise 4.12. This set of sub-normalised states form the *natural variables* of the problem, and in these natural variables, the numerator is linear. It turns out that in any problem involving ensembles, it is always possible to *linearise* this aspect of the problem, by specifying the ensemble in terms of sub-normalised states.

Moving on to the denominator, this is non-linear in a less trivial way. We can however linearise it by *introducing an additional variable* r, subject to the constraints

$$q(i) \leqslant r \quad i = 1, \ldots, o. \tag{4.46}$$

This ensures that $\max_i q(i) \leqslant r$. Indeed, one way to view this is that $\max_i q(i) = \|\vec{q}\|_\infty \leqslant r$, i.e. the ℓ_∞ norm of the vector $\vec{q}$ with elements $q_i = q(i)$ is bounded by r.[4] What we have done is to make use of the same idea for converting the ℓ_∞ norm into an LP, as done in (1.7). We can therefore consider the problem

$$\text{maximise} \quad \frac{\sum_i \operatorname{tr}(M_i \sigma_i)}{r} \tag{4.47a}$$

$$\text{subject to} \quad \sigma_i \succcurlyeq 0 \qquad i = 1, \ldots, o, \tag{4.47b}$$

$$\sum_i \operatorname{tr}(\sigma_i) = 1, \tag{4.47c}$$

$$\operatorname{tr}(\sigma_i) \leqslant r \qquad i = 1, \ldots, o. \tag{4.47d}$$

Similar to previously, it can be shown that the optimal value of this problem is equal to the optimal value of the original problem (4.33), by carefully using the optimal variables for each problem to define variables for the other, and showing that this leads to both upper and lower bounds. This is carried out explicitly in exercise 4.13.

The form (4.47) is still a non-linear optimisation problem, however we have reduced the non-linearity to only the *fractional form*. Both the numerator and the denominator are now *linear* functions of the variables of the problem. It happens that such *linear-fractional* optimisation problems can *always* be re-cast as semidefinite programs, using a standard technique. In exercise 4.14, we consider the general instance of this problem, and show that it can always be re-cast as an explicit SDP, which in our case, leads directly to (4.34).

Exercises

4.11 In this exercise we will find the optimal ensemble for (4.31), and show that it has a trivial form—involving only a single state.

(a) Consider a partial optimisation over only the variables ρ_i. Show that

$$P^*_{\text{guess}} = \text{maximise} \quad \sum_i q(i) \|M_i\|_\infty \tag{4.48}$$

with associated optimal variables ρ_i^* being rank-1 projectors onto eigenvectors of M_i with maximum eigenvalue.

(b) Now optimise over $q(i)$ to show that

$$P^*_{\text{guess}} = \max_i \|M_i\|_\infty, \tag{4.49}$$

[4] Note that since $\vec{q} \geqslant 0$, we do not need to include here the lower bound $-r \leqslant q(i)$ that would otherwise be needed for the ℓ_∞ norm.

with associated optimal variable $q(i)$ equal to 1 for any single j such that $\|M_j\|_\infty = \max_i \|M_i\|_\infty$, and all other $q(i) = 0$.

(c) Consider now the case that for all a, $\|M_a\|_\infty < 1$. Explain in this case, using the structure of the discrimination game, why there is a better guessing strategy than using the measurement $\mathbb{M}$.

This shows that we should consider a more general guessing strategy in order to analyse this game properly, whereby we don't force ourselves to use the outcome of the measurement as the guess directly, but consider also more general ways of producing guesses from the measurement result. Since we are ultimately not really interested in this (trivial) problem, we will not delve into the details of this further, but note that this is another way to see why this is not an interesting problem to consider.

4.12 Consider an ensemble of quantum states, $\mathcal{E} = \{q(i), \rho_i\}$ and an associated collection of sub-normalised states $\tilde{\mathcal{E}} = \{\sigma_i\}$, where $\sigma_i = q(i)\rho_i$. Show that these are two equivalent specifications of the same ensemble. More precisely, show that there is a unique mapping back from $\tilde{\mathcal{E}}$ to $\mathcal{E}$, which maps each sub-normalised state σ_i into the pair $(q(i), \rho_i)$.

4.13 In this exercise we will show that the optimal value of (4.47) is equal to the optimal value of (4.33), in a similar fashion to the presentation in the main text.

(a) Consider an optimal ensemble for (4.33), $\mathcal{E}^* = \{q^*(i), \rho_i^*\}$. Defining variables σ_i and r for (4.47) via

$$\sigma_i = q^*(i)\rho_i^*, \qquad r = \max_j q^*(j), \tag{4.50}$$

show that these variables are feasible, and hence that the optimal value α^* of (4.33) provides a lower bound on the optimal value $\hat{\alpha}^*$ of (4.47), $\alpha^* \leqslant \hat{\alpha}^*$.

(b) Now consider an optimal set of variables for (4.47), $\{\sigma_i^*\}$ and r^*. Defining variables for (4.33) via

$$q(i) = \mathrm{tr}(\sigma_i^*), \qquad \rho_i = \frac{\sigma_i^*}{\mathrm{tr}(\sigma_i^*)}, \tag{4.51}$$

show that these variables are feasible, and hence that the optimal value $\hat{\alpha}^*$ of (4.47) provides a lower bound on the optimal value α^* of (4.33), $\hat{\alpha}^* \leqslant \alpha^*$. *Together (a) and (b) show that $\alpha^* = \hat{\alpha}^*$, and so the two optimisation problems* (4.33) *and* (4.47) *are equivalent, and both evaluate to the optimal advantage provided by* $\mathbb{M}$ *in a quantum state discrimination task*.

4.14 Consider a general type of *linear-fractional* optimisation problem

$$\text{maximise} \quad \frac{\mathrm{tr}(AX) + a}{\mathrm{tr}(DX) + d} \tag{4.52a}$$

$$\text{subject to} \quad \Phi_i(X) = B_i \qquad i = 1, \ldots, m, \tag{4.52b}$$

$$\Gamma_j(X) \preccurlyeq C_j \qquad j = 1 \ldots, n, \tag{4.52c}$$

$$X \succcurlyeq 0, \tag{4.52d}$$

which has an identical set of constraints (and therefore feasible set) as the general SDP from (2.1), however the objective function is the fraction of two linear functions. To this problem we associate the following SDP

$$\text{maximise} \quad \operatorname{tr}(AS) + at \tag{4.53a}$$

$$\text{subject to} \quad \Phi_i(S) = B_i t \qquad i = 1, \dots, m, \tag{4.53b}$$

$$\Gamma_j(S) \preccurlyeq C_j t \qquad j = 1\dots, n, \tag{4.53c}$$

$$S \succcurlyeq 0, \quad t \geqslant 0, \tag{4.53d}$$

$$\operatorname{tr}(DS) + dt = 1. \tag{4.53e}$$

We will show in this exercise that the optimal value of these two problems coincide, and hence (4.53) can be viewed as the SDP formulation of the original non-linear optimisation problem (4.52).

(a) Consider an optimal variable X^* for the original problem (4.52), and potential variables for the SDP (4.53)

$$S = \frac{X^*}{\operatorname{tr}(DX^*) + d}, \qquad t = \frac{1}{\operatorname{tr}(DX^*) + d}. \tag{4.54}$$

Show that these variables are feasible for the SDP (4.53) and hence that the optimal value α^* of the original problem (4.52) is a lower bound on the optimal value $\tilde{\alpha}^*$ of the SDP (4.53), $\alpha^* \leqslant \tilde{\alpha}^*$.

(b) Conversely, consider now an optimal set of variables S^* and t^* for the SDP (4.53) and assume first that $t^* > 0$. Defining a potential variable X for the original problem (4.52) via

$$X = \frac{S^*}{t^*}, \tag{4.55}$$

show that this is feasible and hence that $\tilde{\alpha}^* \leqslant \alpha^*$, i.e. that the optimal value of the SDP (4.53) lower bounds the optimal value of the original problem (4.52).

(c) We will now consider the edge case, not covered in part (b), that $t^* = 0$. Let us pick a feasible variable for the original problem (4.52), and denote it X_0. Show that the potential variable

$$X = X_0 + vS^* \tag{4.56}$$

is feasible for the SDP (4.53) whenever $v \geqslant 0$. Furthermore, show that in the limit $v \to \infty$, the objective function of the original problem (4.52) becomes

$$\lim_{v\to\infty} \frac{\operatorname{tr}[A(X_0 + vS^*)] + a}{\operatorname{tr}[D(X_0 + vS^*)] + d} = \tilde{\alpha}^*. \tag{4.57}$$

This shows that, even if $t^* = 0$, we can nevertheless find a feasible variable for the original problem (4.52) *which achieves a value as close to the optimal value $\tilde{\alpha}^*$ of the SDP* (4.53), *as desired. Together with part (b), this shows that the SDP* (4.53) *is an equivalent problem to the original problem* (4.52), *allowing us to recover both the optimal value (potentially as a limit), and the optimal variables.*

4.3.2 Generalised robustness of measurement informativeness

In the previous section it was seen that when considering a fixed measurement, it is possible to quantify the advantage that it provides in any quantum state discrimination game via solving the SDP (4.34). This was so because it was shown that the SDP is equivalent to (4.33), the original form of the problem. We will now show that *duality* provides further insight into this task, and that the advantage offered by a measurement is captured by a geometrical quantifier known as the *generalised robustness of measurement informativeness*.

Our starting point, as before, is to write down the Lagrangian associated to the primal SDP, in this case as given in (4.34). It is

$$\mathcal{L} = \sum_i \mathrm{tr}(M_i \omega_i) + \sum_i \mathrm{tr}(Y_i \omega_i) + \sum_i z_i[1 - \mathrm{tr}(\omega_i)] \tag{4.58a}$$

$$= \sum_i \mathrm{tr}[\omega_i(M_i + Y_i - z_i \mathbb{1})] + \sum_i z_i, \tag{4.58b}$$

where we have introduced two sets of dual variables, $\{Y_i\}$ and $\{z_i\}$, associated with the left-hand and right-hand constraints in (4.34b) respectively. The Lagrangian upper bounds the primal objective value for all feasible $\{\omega_i\}$ if $Y_i \succcurlyeq 0$ and $z_i \geqslant 0$ for $i = 1, \ldots o$, and becomes independent of the primal variables when $M_i + Y_i - z_i \mathbb{1} = 0$. This leads us directly to the dual SDP

$$\text{minimise} \quad \sum_i z_i \tag{4.59a}$$

$$\text{subject to} \quad M_i + Y_i = z_i \mathbb{1} \qquad i = 1, \ldots, o \tag{4.59b}$$

$$Y_i \succcurlyeq 0, \quad z_i \geqslant 0 \qquad i = 1, \ldots, o. \tag{4.59c}$$

In exercise 4.15 we show that both the primal (4.34) and dual (4.59) are strictly feasible, and hence strong duality holds. The dual SDP (4.59) thus also evaluates to the advantage provided by the measurement $\mathbb{M}$. Interestingly, we can interpret the dual in a geometrical way, by carefully analysing the nature of the dual variables.

First of all, since Y_a is added to M_a,[5] we might naturally be led to wonder if it can be viewed itself as a *measurement*. Since each Y_a is positive semidefinite, they satisfy the first requirement to be interpreted as measurement operators. Considering their normalisation, we see, by summing (4.59b) over a, that

$$\sum_a M_a + \sum_a Y_a = \sum_a z_a \mathbb{1} \tag{4.60a}$$

$$\Rightarrow \sum_a Y_a = \left(\sum_a z_a - 1\right)\mathbb{1}. \tag{4.60b}$$

Although the Y_a are not themselves normalised, since their sum is *proportional* to the identity operator, we can nevertheless extract a normalised measurement from them. In particular, by defining

[5] Note that we switch index from i to a at this stage, as we tend to think of a as the outcome of the measurement, and i as labelling a state from an ensemble. It is therefore nice to change the (dummy) label, to emphasise that the focus is now back on the measurement.

$$r = \sum_a z_a - 1, \qquad N_a = \frac{1}{r} Y_a, \tag{4.61}$$

then $\mathbb{N} = \{N_a\}$ forms a valid measurement. Note that from (4.60b) we see that $r \geqslant 0$, since the sum of a collection of positive semidefinite operators is always positive semidefinite, and it would therefore be a contradiction if the right-hand side, equal to $r\mathbb{1}$, were not positive semidefinite, which is the case whenever $r < 0$.

In a similar fashion, we can also extract a normalised probability distribution from the dual variables z_a. In particular,

$$p(a) = \frac{z_a}{1 + r}, \tag{4.62}$$

is readily seen to form a normalised set of probabilities. We can therefore re-express (4.59) in terms of the variables N_a, $p(a)$ and r as

$$\text{minimise} \quad 1 + r \tag{4.63a}$$

$$\text{subject to} \quad \frac{M_a + rN_a}{1 + r} = p(a)\mathbb{1} \qquad a = 1, \ldots, o, \tag{4.63b}$$

$$N_a \geqslant 0, \qquad \sum_a N_a = \mathbb{1}, \tag{4.63c}$$

$$p(a) \geqslant 0, \qquad \sum_a p(a) = 1, \tag{4.63d}$$

$$r \geqslant 0. \tag{4.63e}$$

In fact, we have not quite shown this reformulation is equivalent, only that it provides a lower bound—since it could be the case that we have actually relaxed the problem by optimising over all measurements $\mathbb{N}$ and probabilities $p(a)$. However, in exercise 4.16 it is shown that the two problems are indeed equivalent. In this new form the problem (4.59) is now expressed in a *non-linear* form (since the variables appear multiplied together in the constraint (4.63b)). The benefit of re-expressing it this way is because this form allows us to see the geometrical nature of the problem, and this geometrical nature arises in many contexts. This type of optimisation is known as a *generalised robustness*.

In (4.63) we will interpret the measurement $\mathbb{N} = \{N_a\}$ as a type of *generalised noise* that is added to our measurement of interest, $\mathbb{M} = \{M_a\}$. We imagine that instead of performing the measurement $\mathbb{M}$ all of the time, with probability $r/(1 + r)$ the measurement $\mathbb{N}$ is instead performed, which is thought of as *degrading* our measurement[6]. Why is this viewed as a degradation? We can similarly interpret the right-hand side of (4.63b) as constituting a *trivial* or *useless* measurement,

[6] We warn the reader that the parameterisation $(p, 1 - p)$, with $0 \leqslant p \leqslant 1$, can also be found instead of $(1/(1 + r), r/(1 + r))$ in the literature about this topic. Here we will stick to the later because this parameterisation appears more directly from the duality theory.

$\mathbb{T} = \{T_a\} = \{p(a)\mathbb{1}\}$. This measurement is trivial as for all quantum states it leads to the same measurement outcomes,

$$\mathrm{tr}(T_a\rho) = p(a) \quad \text{independent of } \rho \tag{4.64}$$

Such a measurement doesn't actually need to be performed at all; a classical random variable distributed according to $p(a)$ can be announced as the measurement outcome, rather than performing the measurement. We therefore see why the measurement $\mathbb{N}$ is considered as noise: this noise is mixed with the measurement $\mathbb{M}$ until the measurement becomes completely useless or trivial.

We therefore see that the optimisation (4.63) seeks to find the *worst-case* noise $\mathbb{N}$, such that when this noise is mixed with the *smallest probability* $r/(1 + r)$, the resulting measurement becomes useless. Said the other way, having solved the problem (4.63), we are guaranteed that if the probability of mixing in noise is smaller than $r^*/(1 + r^*)$, then definitely the measurement hasn't become trivial yet.

All of this can be most easily conveyed using a diagram. As depicted in figure 4.1, we can represent the space of all possible measurements $\mathcal{M}$ (of a fixed dimension and with a fixed number of outcomes)

$$\mathcal{M} = \{\mathbb{M} \mid M_a \succcurlyeq 0,\ a = 1, \ldots, o,\ \sum_a M_a = \mathbb{1}\}, \tag{4.65}$$

which is a convex set. The measurement of interest $\mathbb{M}$ is a point in this space. The set of all trivial measurements,

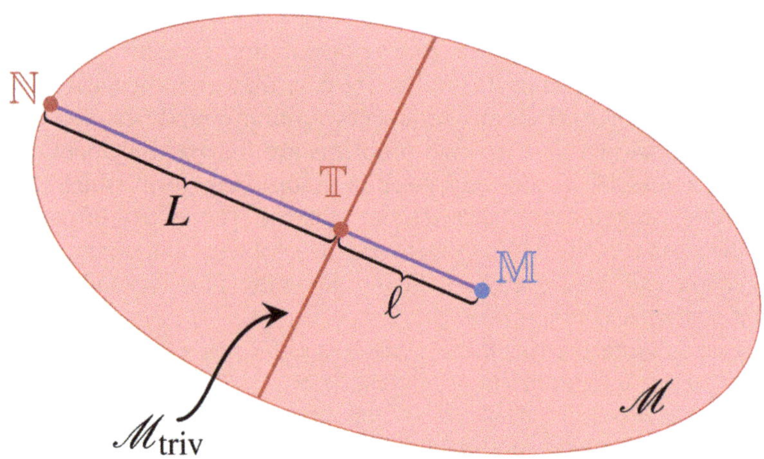

Figure 4.1. *Geometric interpretation of generalised robustness.* Simple representation of the space of measurements $\mathcal{M}$. The subset of trivial measurements, which is lower dimensional, is represented here as a line segment. The noise measurement $\mathbb{N}$ sits on the other side of the trivial set compared to $\mathbb{M}$, and the line joining them intersects this set at a trivial measurement $\mathbb{T}$. The generalised robustness is the ratio of the distance ℓ between $\mathbb{M}$ and $\mathbb{T}$ and the distance L between $\mathbb{N}$ and $\mathbb{T}$.

$$\mathcal{M}_{\text{triv}} = \left\{ \mathbb{M} \mid M_a = p(a)\mathbb{1},\ p(a) \geqslant 0,\ \sum_a p(a) = 1 \right\} \tag{4.66}$$

forms a *convex subset* of $\mathcal{M}$, which is of lower dimension. Picking a noise measurement $\mathbb{N}$, the line segment between this measurement and $\mathbb{M}$ can be drawn. We can see that only those $\mathbb{N}$ such that this line segment intersects the set of trivial measurements $\mathcal{M}_{\text{triv}}$ should be considered. We can denote the two lengths—between $\mathbb{M}$ and the point of intersection, and between $\mathbb{N}$ and the point of intersection—by ℓ and L respectively. In terms of these lengths we have

$$r = \frac{\ell}{L} \tag{4.67}$$

that is, geometrically, r has the interpretation of being the relative distance of $\mathbb{M}$ from being a trivial measurement, compared to the distance of the noise $\mathbb{N}$ from being trivial. The optimisation (4.63) therefore seeks to minimise this ratio—i.e. to find a noise measurement $\mathbb{N}$ that is the furthest from being trivial compared to $\mathbb{M}$. One particularly nice aspect about this geometrical understanding is that if we *fix a direction*, we fix ℓ. We can therefore always find a better r by making L larger—that is, by moving $\mathbb{N}$ further away, in the same direction. This will always be possible until the *boundary* of the set is reached, meaning that the optimal noise measurement $\mathbb{N}^*$ will always be an *extremal* measurement. As we will see below, this can be viewed as a geometrical manifestation of *complementary slackness*, which provides a number of key insights into the structure of the optimal ensemble $\mathcal{E}^*=\{q^*(i),\ \rho_i^*\}$ and optimal noise measurement $\mathbb{N}^*=\{N_a^*\}$.

4.3.3 Structure of the optimal ensemble and noise measurement

We will now see that a significant amount can be learnt about the structure of the optimal ensemble and noise measurement, using the complementary slackness conditions, in conjunction with the other facts already learnt to date.

From the Lagrangian (4.58), we can readily write down the complementary slackness conditions for our discrimination problem. Namely, the optimal primal and dual variables satisfy

$$Y_i^* \omega_i^* = 0, \qquad z_i^*[1 - \operatorname{tr}(\omega_i^*)] = 0, \qquad i = 1, \ldots, o \tag{4.68}$$

We already showed previously that $\operatorname{tr}(\omega_i^*) = 1$ without loss of generality. In exercise 4.17 below, it will also be shown that $z_i^* > 0$ for all i, hence from the second complementary slackness condition in (4.68), a second, independent way of arriving at the same conclusion is obtained. In (4.38) we saw how to use ω_i^* in order to generate an ensemble, which is moreover an *optimal* ensemble. Since we now know that $\operatorname{tr}(\omega_i^*) = 1$, it follows that

$$q^*(i) = \frac{1}{o}, \qquad \rho_i^* = \omega_i^*, \tag{4.69}$$

which shows that the optimal ensemble sends the states ρ_i^* *uniformly at random* with probability $1/o$. We also saw in (4.61) how to use Y_a^* in order to generate an optimal noise measurement, $N_a^* = Y_a^*/r^*$. Together, this allows us to re-express the first complementary slackness condition in (4.68) as

$$N_i^*\rho_i^* = 0 \qquad i = 1, \ldots, o. \tag{4.70}$$

To see what this is means, consider playing the quantum state discrimination game for the ensemble $\mathcal{E}^* = \{q^*(i), \rho_i^*\}$ with the measurement $\mathbb{N}^*=\{N_a^*\}$. The average success probability is

$$P_{\text{guess}} = \sum_i q^*(i)\text{tr}(N_i^*\rho_i^*) = 0. \tag{4.71}$$

That is, this is the *worst* possible measurement that could be used, which fails to correctly identify the state ever! Mathematically, (4.70) shows that N_a^* has at least one zero eigenvalue, and is therefore an extremal measurement, as we saw above geometrically.

Finally, we can also see that if none of the measurement operators M_a are trivial —in the sense of being proportional to the identity—then the optimal ensemble cannot contain full-rank states. Indeed, from (4.70), if ρ_i^* were full rank, then it would immediately imply that $N_i^* = 0$. This in turn would mean that no noise is added to M_i in order for it to become trivial. However, by assumption this was not the case, and so we can infer, since $N_i^* \neq 0$, then ρ^* is not full rank.

Exercises

4.15 Show that both the primal SDP (4.34) and dual SDP (4.59) are strictly feasible, and hence that strong duality holds.

4.16 In this exercise we will show that the two problems (4.59) and (4.63) are equivalent.
- (a) Assume an optimal set of variables for (4.59). Show that, through the definitions (4.61) and (4.62), that the optimal value of (4.59) upper bounds the optimal value of (4.63).
- (b) Assume now an optimal set of variables for (4.63). Using these to define a feasible set of variables for (4.59), show that the optimal value of (4.63) upper bounds the optimal value to (4.59).

  This shows that the two problems have the same optimal value, and are therefore equivalent.

4.17 In this exercise we show that, just as the primal SDP (4.34) could be solved analytically, so too can the dual SDP (4.59).

(a) Show that in (4.59) Y_a are slack variables and that the an equivalent form of the dual is

$$\text{minimise} \quad \sum_i z_i \tag{4.72a}$$

$$\text{subject to} \quad z_i \mathbb{1} \geqslant M_i \qquad i = 1, \ldots, o, \tag{4.72b}$$

$$z_i \geqslant 0, \qquad i = 1, \ldots o. \tag{4.72c}$$

(b) Assume that the constraint $z_a \geqslant 0$ is saturated for some a, i.e. that $z_a^* = 0$. Explain why this implies that the corresponding $M_a = 0$. *Since, by the nature of the problem,* $M_a \neq 0$, *this shows that* $z_a > 0$ *without loss of generality.*

(c) Given part (b), use the dual formulation of the maximum eigenvalue SDP from (2.28) to analytically solve (4.72), and confirm that it agrees with the analytic solution to the primal problem from (4.42).

(d) Starting from the Lagrangian (4.58), write down the complementary slackness conditions for this problem.

(e) Use complementary slackness, and part (b) to provide an alternative proof that $\text{tr}(\omega_i^*) = 1$.

4.4 Concluding remarks

In this chapter we considered problems involving quantum measurements. As we have seen, optimising over quantum measurements also fits perfectly inside the SDP framework, since they are defined by semidefinite constraints. We have also studied the problem of quantum state discrimination from two alternative points of view, and seen numerous applications of semidefinite programming in this context. The main ideas that we hope you learnt in this chapter are:

- **Measurement estimation.** Similarly to quantum state estimation, the problem of estimating which measurement is being performed given a set of observed outcomes can be cast as an SDP. Moreover, we can consider the case of finite statistics, and quantify how far from being feasible a set of measurement data is—equation (4.3).
- **Minimum-error quantum state discrimination.** The guessing probability of minimum-error quantum state discrimination can be cast as an SDP, allowing for the optimal measurements for this task to be found—(4.9).
- **Operational interpretation of the trace distance.** The trace distance quantifies how well two states can be distinguished in symmetric and binary minimum-error binary quantum state discrimination—see (4.13).
- **Optimality of measurements.** Complementary slackness allows us to find necessary and sufficient conditions for the optimality of measurements for minimum-error quantum state discrimination—(4.16).

- **Unambiguous state discrimination.** We can determine the set of measurements that minimise the probability of failure in unambiguous quantum state discrimination using SDPs—(4.29).
- **Linear-fractional optimisation problems.** We have seen how to transform an optimisation problem were the objective function involves the division of two linear functions into an SDP—see exercise 4.14.

4.5 Further reading

- Bae J and Kwek L-C 2015 Quantum state discrimination and its applications *J. Phys. A: Math. Theor.* **48** 083001

4.6 Advanced topics

4.6.1 Unambiguous quantum state discrimination revisited

In this section we will return to unambiguous state discrimination. We will begin by analysing more carefully when unambiguous state discrimination is impossible. We will then use this insight to see how we can interpret the dual formulation in terms of approximate linear dependence.

Suppose that the ensemble of states—viewed as a collection of sub-normalised states—is *linearly dependent*, meaning that it is possible to find a set of real numbers μ_i for $i = 1, \dots, m$ such that

$$\sum_i \mu_i q(i)\rho_i = 0. \tag{4.73}$$

The advantage of using again sub-normalised states, similar to the reasoning for imposing (4.28), is that it ensures that we only consider states that occur with non-zero probability, which will be advantageous below.

After multiplying (4.73) by M_a, for some fixed value of a, and taking the trace, we see, given that the measurement satisfies $q(i)\mathrm{tr}(M_a\rho_i) = 0$ for all values of $i \neq a$, that all but the term $i = a$ vanishes, and therefore we have

$$q(a)\mu_a \mathrm{tr}(M_a\rho_a) = 0. \tag{4.74}$$

Therefore as long as $\mu_a \neq 0$, it must be the case that $p(a, i = a) = q(a)\mathrm{tr}(M_a\rho_a) = 0$. That is, the measurement can never unambiguously identify the state ρ_a. In such a case, the 'failure outcome' $a = \varnothing$ is returned all of the time, indicating that we are never certain which state has been sent.

When will we have that none of the μ_i vanish? This will occur when the set of sub-normalised states $\{q(i)\rho_i\}$ is linearly dependent as a whole, but no subset is linearly dependent (i.e. all subsets are linearly independent).

We will now see why this insight into when unambiguous state discrimination is impossible is useful, and can be extended to understand how well unambiguous state discrimination can be performed in general. In particular, by considering the dual formulation of (4.29), we will see in what follows that the average probability of success in unambiguous quantum state discrimination can be lower bounded by quantifying *how far the set of sub-normalised states is from being linearly dependent.*

In exercise 4.18, it is shown that the dual formulation of (4.29) is

$$P^*_{\text{guess}} = \text{minimise} \quad \operatorname{tr}(Z) \tag{4.75a}$$

$$\text{subject to} \quad Z \succcurlyeq q(a)\rho_a + \sum_{i \neq a} \mu_{ai} q(i)\rho_i \qquad a = 1, \ldots, m, \tag{4.75b}$$

$$Z \succcurlyeq 0. \tag{4.75c}$$

This SDP has a structure that is somewhat familiar. In particular, there is some similarity to the SDP for the trace norm, given in (2.8), except the right-hand side of (4.75b) contains the dual variables μ_{ai}. This suggests that one way to proceed is via *partial optimisation*—to consider first that the dual variables μ_{ai} are *fixed*—and analysing the structure of the dual problem only from the perspective of the dual variable Z, before returning to the problem of optimising the remaining variables. This leads us to consider the following SDP,

$$\text{minimise} \quad \operatorname{tr}(Z) \tag{4.76a}$$

$$\text{subject to} \quad Z \succcurlyeq X, \tag{4.76b}$$

$$Z \succcurlyeq 0, \tag{4.76c}$$

which we arrive at by considering only a single constraint from (4.75b), and treating the right-hand side as (fixed) data X. In exercise 4.19, we show that this SDP evaluates exactly to $\|X^{(+)}\|_1$, the trace norm of the positive part of the operator X.

How can we use this to understand (4.75)? First of all, there is not a single operator X, but one for each value of a, and these are not data, but rather variables, so we will define $X_a = q(a)\rho_a + \sum_{i \neq a} \mu_{ai} q(i)\rho_i$, which are the operators appearing on the right-hand side of (4.75b). Second, if there was only a single X_a, then the above shows that (4.75) would evaluate to $\| X_a{}^{(+)}\|_1$. The fact there are multiple X_a, but only a single variable Z appearing in all of the constraints (4.75b) is similar to what happens in the LP for the ℓ_∞ norm (1.7)—where a scalar variable bounded every component of a vector. We see immediately that $\operatorname{tr}(Z)$ definitely cannot be smaller than $\| X_a{}^{(+)}\|_1$ for any value of a. However, it may furthermore not even be possible to achieve any of these values, since there will in general be overlap between the positive parts of the $X_a{}^{(+)}$ for different values of a. We can nevertheless obtain the interesting *lower bound* on the guessing probability by ignoring this fact,

$$P^*_{\text{guess}} \succcurlyeq \text{minimise} \quad \max_a \| X_a{}^{(+)}\|_1 \tag{4.77a}$$

$$\text{subject to} \quad X_a = q(a)\rho_a + \sum_{i \neq a} \mu_{ai} q(i)\rho_i \qquad a = 1, \ldots, m. \tag{4.77b}$$

Let us now try to understand why this lower bound—which at first sight might look less informative—is in fact a useful way of understanding the dual.

First of all, notice that if we take the uninteresting case—where the set of unnormalised states $\{q(i)\rho_i\}$ is linearly dependent, then each X_a can be made to vanish. In particular, we can take

$$\mu_{ai} = \frac{\mu_i}{\mu_a} \tag{4.78}$$

where the μ_i are from (4.73). Direct substitution shows that this makes X_a vanish, in which case $\|X_a^{(+)}\|_1$ also vanishes for all a, and so $P^*_{\text{guess}} = 0$ from the dual (4.75), before even getting to the lower bound.

Let us therefore now assume that we are in the case of interest—where the set of unnormalised states is linearly independent. We may then imagine looking for a set of numbers μ_i such that $\sum_i \mu_i q(i)\rho_i$ is '*smallest*'. However, what exactly is meant by smallest is a subtle question. The first difficulty encountered is that there is an ambiguity in defining μ_i, since we can always consider a second choice $\mu_i' = \gamma\mu_i$, for some choice of γ, and this should not change our notion of 'smallness'. That is, to start with we need to fix a type of *normalisation* or *scale* for the μ_i.

With this in mind, we can now see that each X_a in (4.77b) can be viewed as *making a specific choice for the normalisation* of a set of μ_i. In particular, X_a fixes $\mu_a = 1$, and sets $\mu_i = \mu_{ai}$ for $i \neq a$. This is a somewhat natural normalisation condition, which can be seen to fix an overall scale for the μ_i. There is no good reason to single out a specific μ_a to set to be equal to one—since all are equivalent as far as the problem is concerned. We will see shortly how this is dealt with.

Let us first assume that we have fixed a normalisation convention—$\mu_a = 1$ for some specific choice of a. We can choose to measure how far the resulting operator X_a is from vanishing using $\|X_a^{(+)}\|_1$, i.e. by the trace norm of its positive part. Each choice for a in principle could lead to a different number however. One natural choice is to consider the *worst case*, i.e. the largest distance, as a figure of merit.

Mathematically, this is precisely $\max_a \|X_a^{(+)}\|_1$, which we now understand can be viewed as quantifying how far the set of unnormalised states $q(i)\rho_i$ is from being linearly dependent, irrespective of which μ_i is fixed to be one. By optimising over all choices of μ_{ai}, we seek to find the best set of coefficients which make the combination *approximately* vanish.

The lower bound optimisation problem (4.77) can thus be interpreted as calculating the distance of the set of unnormalised states $\{q(i)\rho_i\}$ from being linearly independent. We see therefore that how close a set of states is to being linearly independent in a crucial property which bounds how well the set can be unambiguously discriminated.

Exercises

4.18 In this question we will derive the dual SDP formulation of unambiguous state discrimination, as given in (4.75).
  (a) Write down the Lagrangian associated to the primal SDP (4.29).
  (b) Identify the constraints satisfied by the dual variables: (i) in order for the Lagrangian to upper bound the primal value; (ii) in order for the Lagrangian to be independent of the primal variables.
  (c) Show that part (b) leads to the dual SDP stated in (4.75).

4.19 In this question we will show that the SDP (4.76) evaluates to $\|A^{(+)}\|_1$.

(a) Show that $Z = A^{(+)}$ achieves the value $\|A^{(+)}\|_1$ for the SDP (4.76). *This shows that $\|A^{(+)}\|_1$ is an upper bound on the value of the SDP.*

(b) Show that the dual SDP of (4.76) is

$$\text{maximise} \quad \mathrm{tr}(AX) \tag{4.79a}$$

$$\text{subject to} \quad X + Y = \mathbb{1}, \tag{4.79b}$$

$$X \succcurlyeq 0, \quad Y \succcurlyeq 0. \tag{4.79c}$$

(c) Verify that strong duality holds for this primal-dual SDP pair.

(d) Pick a pair of dual variables X and Y such that the value of the dual is $\|A^{(+)}\|_1$. *Since this a lower bound on the primal SDP, this guarantees. that the SDP evaluates to $\|A^{(+)}\|_1$.*

4.20 Show that if the set of states $\{\rho_i\}$ is linearly dependent then the value of the dual SDP in its original form, (4.75), is zero, i.e. find a feasible Z such that $\mathrm{tr}(Z) = 0$.

Paul Skrzypczyk and Daniel Cavalcanti

Chapter 5

Quantum entanglement

In this chapter we will turn our attention to *quantum entanglement*—one of the most important aspects of quantum theory from the perspective of quantum information. As we will see, the techniques of semidefinite programming play a wide ranging role in entanglement theory, starting from one of the simplest criteria for detecting entanglement, known as the *positive-partial-transpose criteria*. This leads to a natural way to quantify entanglement, through the so-called *negativity*. In general, the set of unentangled—or separable—states has a complicated structure, which has no computationally easy characterisation. However, semidefinite programs (SDPs) can be used to approximate the set of separable states, and allows for bounds to be obtained on many quantities of interest.

We will also introduce a new and powerful idea in this chapter—that of a sequence or *hierarchy of SDP criteria*, which in the limit converge to the set of separable states, known as *k-symmetric extensions*. The idea of a hierarchy is a general technique, and this section aims to demonstrate how this powerful idea works.

5.1 Entanglement of pure and mixed states

A pure bipartite $d_A \times d_B$ quantum state $|\Psi\rangle_{AB}$ is entangled if and only if it cannot be factorised, i.e. written as a direct product state

$$|\Psi\rangle_{AB} = |\psi\rangle_A|\phi\rangle_B, \tag{5.1}$$

where $|\psi\rangle_A$ is an arbitrary pure state for system A, $|\phi\rangle_B$ is an arbitrary pure state for system B, and we have used the notation $|\psi\rangle_A|\phi\rangle_B$ to denote the tensor product state $|\psi\rangle_A \otimes |\phi\rangle_B$ (this could also be denoted $|\psi, \phi\rangle_{AB}$). There is a straightforward way to check if such a decomposition exists: We can use two facts: (i) that the reduced density operator of a bipartite pure state is mixed if and only if the state is entangled; (ii) only for pure density operators ρ do we have $\mathrm{tr}(\rho^2) = 1$. Together, this shows that

$$|\Psi\rangle_{AB} \text{ is entangled if and only if } \mathrm{tr}(\rho_A^2) < 1,$$

where $\rho_A = \mathrm{tr}_B(|\Psi\rangle_{AB}\langle\Psi|)$.

doi:10.1088/978-0-7503-3343-6ch5

© IOP Publishing Ltd 2023

When we move onto mixed bipartite states, the situation becomes more intricate. A $d_A \times d_B$ quantum state ρ_{AB} is entangled if and only if it cannot be written as a convex combination of product states, namely

$$\rho_{AB} \neq \sum_{\lambda} p(\lambda)\rho_{\lambda}^{A} \otimes \rho_{\lambda}^{B} \tag{5.2}$$

where $\lambda = 1, \dots, N$, is a classical random variable taking on N values, with probabilities $p(\lambda)$, and $\{\rho_{\lambda}^{A}\}$ and $\{\rho_{\lambda}^{B}\}$ are arbitrary sets of states for system A and B respectively, which need not be related in any way. Any state which has a decomposition of the form (5.2) is said to be *separable*.

One of the most basic questions in the theory of entanglement is to determine whether a generic density operator ρ_{AB} is entangled or not. Unlike in the special case of pure states, in general this is known to be a hard problem, for which no efficient algorithm should exist, for reasons of computational complexity (in particular, the entanglement problem has been shown to be an NP-hard problem). Intuitively, from (5.2), we see that to answer the entanglement question requires searching over infinitely many decompositions of a state. Unlike for problems previously encountered in this book, there is no known way around this complexity, and deciding if a quantum state is entangled or not is not currently known to be an SDP. Nevertheless, as will be seen below, progress can still be made on this important question. First, there is an exception to the above rule, when considering small dimensions (to be more specific, if $d_A \times d_B = 4, 6$). In this case, it is in fact possible to determine whether a state is entangled or not using an SDP. Second, we will introduce an important idea: a hierarchy of feasibility semidefinite programs which must all be satisfied if a state is separable. This hierarchy provides a sequence of (outer) approximations to the set of separable states, which can be used in practice to obtain useful and interesting bounds on entanglement.

5.2 The positive-partial-transpose criterion

Although it is computationally difficult to build a test for entanglement that works for quantum states of any dimension, this does not rule out the possibility of finding a simple criterion that certifies the entanglement of some, but not all, quantum states. That is, it is very useful to have *sufficient* criteria which guarantee that a state is entangled.

The most famous such criterion is the *positive-partial-transpose* criterion. It is based on the partial transpose linear map, which can be specified as

$$X_{AB}^{\mathrm{T}_A} = \Gamma^{(\mathrm{T}_A)}(X_{AB}) = \sum_{i,j} (|j\rangle\langle i| \otimes \mathbb{I}) X_{AB} (|j\rangle\langle i| \otimes \mathbb{I}) \tag{5.3}$$

where we introduce the notation $\Gamma^{(\mathrm{T}_A)}$ to emphasise the connection with (2.1c), and to highlight that (as we will discuss again later) this is a specific linear map that can be used as a constraint inside an SDP. We can see that the partial transpose simply applies a transposition on the first Hilbert space, while leaving the second Hilbert space untouched.

The positive-partial-transpose criterion says that a bipartite quantum state is necessarily entangled if transposing only one part of it—referred to as a *partial transpose*—turns it into an operator that is not positive semidefinite (and therefore no longer a quantum state). That is,

$$\rho_{AB}^{\mathsf{T}_A} \not\succcurlyeq 0 \Rightarrow \rho_{AB} \neq \sum_{\lambda} p(\lambda)\rho_{\lambda}^{A} \otimes \rho_{\lambda}^{B}. \tag{5.4}$$

This result can be proven easily by contradiction: if a density operator σ_{AB} is separable, then

$$\sigma_{AB}^{\mathsf{T}_A} = \sum_{\lambda} p(\lambda)(\rho_{\lambda}^{A})^{\mathsf{T}} \otimes \rho_{\lambda}^{B} \succcurlyeq 0, \tag{5.5}$$

since $(\rho_{\lambda}^{A})^{\mathsf{T}} \succcurlyeq 0$ whenever $\rho_A \succcurlyeq 0$ (since transposition does not change the eigenvalues of a matrix). Thus, if $\rho_{AB}^{\mathsf{T}_A} \not\succcurlyeq 0$ then ρ_{AB} must necessarily be entangled[1].

The converse of the PPT criterion is not necessarily true: there exist quantum states which are entangled but which also remain positive semidefinite after applying the partial transpose. Such states are called *PPT entangled states*, standing for 'positive-partial-transpose'. When states are not PPT, we refer to them as *NPT entangled states* where NPT stands for 'negative-partial-transpose'.

We can define the set of all PPT states,

$$\mathcal{S}_{\mathrm{PPT}} = \left\{\pi_{AB} \mid \pi_{AB} \succcurlyeq 0,\ \mathrm{tr}(\pi_{AB}) = 1,\ \pi_{AB}^{\mathsf{T}_A} \succcurlyeq 0\right\}, \tag{5.6}$$

and as shown in exercise 5.2, this is a convex set of states. We can similarly define the set of separable states,

$$\mathcal{S}_{\mathrm{sep}} = \Bigg\{\sigma_{AB} \mid \sigma_{AB} \succcurlyeq 0,\ \mathrm{tr}(\sigma_{AB}) = 1,\ \sigma_{AB} = \sum_{\lambda} p(\lambda)\rho_{\lambda}^{A} \otimes \rho_{\lambda}^{B}, \\ \rho_{\lambda}^{A} \succcurlyeq 0,\ \rho_{\lambda}^{B} \succcurlyeq 0,\ \mathrm{tr}(\rho_{\lambda}^{A}) = 1,\ \mathrm{tr}(\rho_{\lambda}^{B}) = 1\Bigg\}. \tag{5.7}$$

The fact that the PPT criterion does not detect all entangled states means that the set $\mathcal{S}_{\mathrm{PPT}}$ of PPT states strictly contains the set $\mathcal{S}_{\mathrm{sep}}$ of separable states, $\mathcal{S}_{\mathrm{sep}} \subset \mathcal{S}_{\mathrm{PPT}}$. We must note however, that there is in fact a set of PPT states for each dimension of Alice's and Bob's system. When considering the smallest possible dimensions, when $d_A = d_B = 2$, $d_A = 2$ and $d_B = 3$ or $d_A = 3$ and $d_B = 2$, then it turns out that the PPT criterion is also a sufficient criterion for a state to be separable, and in these dimensions $\mathcal{S}_{\mathrm{sep}} = \mathcal{S}_{\mathrm{PPT}}$. In these situations, testing whether a quantum state is PPT or not is equivalent of testing whether or not it is separable. This is however the exception to the rule, and in general the two are not equivalent. We can see how the PPT criterion can be used in practice, for the simplest class of entangled states:

[1] We have presented the positive-partial transposition criterion in terms of the transposition with respect to party A, but we could equivalently have used the transposition with respect to party B since $X^{\mathsf{T}_A} = (X^{\mathsf{T}_B})^{\mathsf{T}}$.

Example 5.1 PPT criterion applied to two-qubit isotropic states.
In this simple example we will see how the PPT criterion can be used to detect the entanglement of two-qubit *isotropic states*, mixed states of the form

$$\rho_{AB}(w) = w|\Phi^+\rangle\langle\Phi^+| + (1-w)\frac{\mathbb{1}_A \otimes \mathbb{1}_B}{4}, \tag{5.8}$$

with $|\Phi^+\rangle = \frac{1}{\sqrt{2}}(|0\rangle|0\rangle + |1\rangle|1\rangle)$ the maximally entangled state and $0 \leqslant w \leqslant 1$. Isotropic states are entangled when $w > \frac{1}{3}$ and separable otherwise.

A direct calculation shows that $|\Phi^+\rangle\langle\Phi^+|^{\mathsf{T}_A} = \frac{1}{2}S$, where S is the SWAP unitary operator, $S = \sum_{i,j}|i\rangle|j\rangle\langle j|\langle i|$, such that $S|\psi\rangle|\chi\rangle = |\chi\rangle|\psi\rangle$ for all states $|\psi\rangle$ and $|\chi\rangle$. It then follows that

$$\rho_{AB}(w)^{\mathsf{T}_A} = \frac{w}{2}S + (1-w)\frac{\mathbb{1}_A \otimes \mathbb{1}_B}{4}. \tag{5.9}$$

SWAP has eigenvalues 1 (with multiplicity 3) and -1. The former eigenspace is spanned by the symmetric states $|0\rangle|0\rangle$, $|1\rangle|1\rangle$ and $|\Psi^+\rangle = \frac{1}{\sqrt{2}}(|0\rangle|1\rangle + |1\rangle|0\rangle)$, while the latter corresponds to the antisymmetric maximally entangled state $|\Psi^-\rangle = \frac{1}{\sqrt{2}}(|0\rangle|1\rangle - |1\rangle|0\rangle)$. The eigenvalues of $\rho_{AB}(w)^{\mathsf{T}_A}$ are thus $\frac{1+w}{4}$ (with multiplicity 3) and $\frac{1-3w}{4}$. In order for $\rho_{AB}(w)^{\mathsf{T}_A}$ to be PPT all eigenvalues must be nonnegative, and hence $\frac{1-3w}{4} \geqslant 0$, which is equivalent to $w \leqslant \frac{1}{3}$. Thus, when $w > \frac{1}{3}$ the state is NPT entangled. This thus demonstrates that the PPT criterion is able to detect all entangled two-qubit isotropic states.

In exercise 5.1 you will show that this result generalises to isotropic states of arbitrary dimension.

We end this section by noting that the above can be used to express the problem of whether a given state ρ_{AB} is an NPT entangled state as a feasibility SDP. Namely,

$$\text{find } X \tag{5.10a}$$

$$\text{subject to } X \geqslant 0, \tag{5.10b}$$

$$X = \rho_{AB}^{\mathsf{T}_A}. \tag{5.10c}$$

This simple SDP forces the variable to be equal to the partial transpose of the input quantum state. Only if this operator is positive semidefinite will the problem be feasible. In practice this SDP is too simple to be used—one can simply find the eigenvalues of $\rho_{AB}^{\mathsf{T}_A}$ directly, and determine whether they are nonnegative. It is nevertheless important to realise that the problem is an instance of a feasibility SDP. In what follows, we will go further, and see how to consider an optimisation form of this problem, which will allow us to quantify entanglement.

Exercises

5.1 In this exercise we will show that the PPT criterion is able to detect the entanglement of isotropic states in any dimension. The family of d-dimensional isotropic states is given by

$$\rho_{AB}^{(d)}(w_d) = w_d|\Phi_{(d)}^+\rangle\langle\Phi_{(d)}^+| + (1 - w)\frac{\mathbb{1}_A \otimes \mathbb{1}_B}{d^2}, \tag{5.11}$$

where $|\Phi_{(d)}^+\rangle = \frac{1}{\sqrt{d}}\sum_i|i\rangle|i\rangle$ is the maximally entangled state of two qudits, and where $0 \leqslant w_d \leqslant 1$. d-dimensional isotropic states are entangled when $w_d > \frac{1}{d+1}$ and separable otherwise.

(a) Show that

$$|\Phi_{(d)}^+\rangle\langle\Phi_{(d)}^+|^{\mathsf{T}_A} = \frac{1}{d}S_{(d)},$$

where $S_{(d)} = \sum_{i,j}|i\rangle|j\rangle\langle j|\langle i|$ is the d-dimensional SWAP unitary operator.

(b) Using the fact that the eigenvalues of $S_{(d)}$ are 1 and −1 (with corresponding eigenspaces spanned by symmetric and antisymmetric states, respectively), show that $\rho_{AB}^{(d)}(w_d)$ is NPT whenever it is entangled (i.e. for $w_d > \frac{1}{d+1}$).

This result is more interesting than Example 5.1. *In that example, it was guaranteed that the PPT criterion had to detect the entanglement of the state, since PPT and separable are equivalent for two–qubit states. Here, in contrast, since the states are qudits, it is not guaranteed that the PPT criterion should detect all entangled states in the family. This results shows that it nevertheless does.*

5.3 Entanglement negativity

The PPT criterion can be used to define a quantifier of entanglement, known as *entanglement negativity*. The basic idea behind this quantity is that when a state is more entangled, then there is a sense in which its partial transpose will be a *more negative* operator. As we will see, the negativity of a quantum state can be cast as a semidefinite program. Moreover, through duality we will be able to show that all pure entangled states have a negative-partial transpose.

One measure of 'how negative' an operator is the absolute value of the sum of all of its negative eigenvalues. The larger this number, the 'more negative' the operator is. We can turn the evaluation of this quantity into an SDP by recalling that a (Hermitian) operator A can be decomposed into a positive part $A^{(+)}$ and a negative part $A^{(-)}$,

$$A = A^{(+)} - A^{(-)}, \tag{5.12}$$

where $A^{(+)} \succcurlyeq 0$ and $A^{(-)} \succcurlyeq 0$ (see exercise 2.7). In this form, all of the positive eigenvalues of A (and corresponding eigenprojectors) are collected into $A^{(+)}$, while the (absolute value of the) negative eigenvalues (and corresponding eigenprojectors) are collected into $A^{(-)}$. The utility of writing A in this form is that it provides us with a convenient way to express the absolute value of the sum of the negative eigenvalues of A. Namely, we can re-express this as the trace norm of the negative part, $\|A^{(-)}\|_1$, since this, by definition, is the sum of the eigenvalues of $A^{(-)}$ (which are all positive), which are themselves nothing but the absolute values of the negative eigenvalues of A.

As will be shown in exercise 5.3, $\|A^{(-)}\|_1$ can be cast as the following pair of primal and dual SDPs

$$\begin{aligned} &\text{minimise} \quad \mathrm{tr}(Y) && \text{maximise} \quad -\mathrm{tr}(XA) && (5.13a)\\ &\text{subject to } A = Z - Y, && \text{subject to } \mathbb{1} \geqslant X, && (5.13b)\\ &Y \geqslant 0, \quad Z \geqslant 0. && X \geqslant 0. && (5.13c) \end{aligned}$$

Using these, we can now directly define the *negativity* of a quantum state $\mathcal{N}(\rho_{AB})$ to be the absolute value of the sum of the negative eigenvalues of $\rho_{AB}^{\mathsf{T}_A}$, which can equally be thought of as the trace norm of the negative part of $\rho_{AB}^{\mathsf{T}_A}$, and immediately write it as an SDP in primal form as

$$\mathcal{N}(\rho_{AB}) = \text{minimise} \quad \mathrm{tr}(Y) \tag{5.14a}$$

$$\text{subject to } \rho_{AB}^{\mathsf{T}_A} = Z - Y, \tag{5.14b}$$

$$Y \geqslant 0, \quad Z \geqslant 0, \tag{5.14c}$$

and in dual form as

$$\mathcal{N}(\rho_{AB}) = \text{maximise} \quad -\mathrm{tr}(X^{\mathsf{T}_A}\rho_{AB}) \tag{5.15a}$$

$$\text{subject to } \mathbb{1} \geqslant X, \tag{5.15b}$$

$$X \geqslant 0, \tag{5.15c}$$

where we have used the fact that the adjoint of the partial transpose is the partial transpose itself, i.e. that $\mathrm{tr}(F^{\mathsf{T}_A}G) = \mathrm{tr}(FG^{\mathsf{T}_A})$ for all F and G, in order to slightly re-express the dual objective function.

The dual has an important geometrical interpretation as optimising over *quantitative entanglement witnesses*, or more precisely in this case, quantitative witnesses of NPT entanglement. To elucidate this, we will make a small change of variable,

$$W = X^{\mathsf{T}_A}, \tag{5.16}$$

in terms of which the dual is

$$\mathcal{N}(\rho_{AB}) = \text{maximise} \quad -\mathrm{tr}(W\rho_{AB}) \tag{5.17a}$$

$$\text{subject to } \mathbb{1} \geqslant W^{\mathsf{T}_A}, \tag{5.17b}$$

$$W^{\mathsf{T}_A} \geqslant 0. \tag{5.17c}$$

The operator W is known as an *entanglement witness*. In order to see why, we start from the constraint (5.17c). This constraint implies that $\mathrm{tr}(W\sigma_{AB}) \geqslant 0$ for all separable σ_{AB}, as can be verified directly, using the fact that $\mathrm{tr}(FG) = \mathrm{tr}(F^{\mathrm{T}_A}G^{\mathrm{T}_A})$ for all F and G,

$$\mathrm{tr}(W\sigma_{AB}) = \mathrm{tr}(W^{\mathrm{T}_A}\sigma_{AB}^{\mathrm{T}_A}) \geqslant 0, \tag{5.18}$$

where the inequality follows since $W^{\mathrm{T}_A} \succcurlyeq 0$ from (5.17c), $\sigma_{AB}^{\mathrm{T}_A} \succcurlyeq 0$ as we saw in (5.5), and as we have used many times, the trace of the product of positive semidefinite operators is never negative. What this shows is that the only way in which $\mathrm{tr}(W\rho_{AB})$ can be negative is if ρ_{AB} is entangled. In other words $\mathrm{tr}(W\rho_{AB}) \leqslant 0$ *witness* the entanglement of ρ_{AB}. This is a direct application of the results presented in section 3.1.5, where the dual formulation of an SDP was show to provide a certificate of infeasibility (in the present case, a certificate that the state is not PPT).

The constraint (5.17b) can be understood as setting the normalisation of the entanglement witness. To see this, we consider multiplying the constraint by an arbitrary separable quantum state σ_{AB} and taking the trace, leading to

$$\mathrm{tr}(\sigma_{AB}) \geqslant \mathrm{tr}(W^{\mathrm{T}_A}\sigma_{AB}). \tag{5.19}$$

We can then use the fact that $\mathrm{tr}(W^{\mathrm{T}_A}\sigma_{AB}) = \mathrm{tr}(W\sigma_{AB}^{\mathrm{T}_A})$, and realise that $\sigma_{AB}^{\mathrm{T}_A} = \sigma'_{AB}$ is a separable quantum state (in general different from σ_{AB}). Finally, using the fact that σ_{AB} is normalised, we arrive at the inequality

$$1 \geqslant \mathrm{tr}(W\sigma'_{AB}), \tag{5.20}$$

i.e. that the value that the witness takes on *any* separable state is bounded from above by unity. Altogether, what this shows us is that the constraints in (5.17) enforce that the dual variable W must have a (Hilbert–Schmidt) scalar product with any separable state lying in the interval $0 \leqslant \mathrm{tr}(W\sigma_{AB}) \leqslant 1$. We thus arrive at the realisation that the negativity can be interpreted as the *violation of a quantitative entanglement witness*, that is, by how much the condition $\mathrm{tr}(W\rho_{AB}) \geqslant 0$ is violated. This is summarised graphically in figure 5.1. Although the normalisation $0 \leqslant \mathrm{tr}(W\sigma_{AB}) \leqslant 1$ is rather natural, it not the only choice that could be taken. Interestingly, varying the way in which an entanglement witness is normalised leads to a number of other interesting entanglement quantifiers, as we will see later.

It is also important to note that in the above, the entanglement witnesses are only able to detect entangled states that have a negative-partial transpose. Any entangled state that has a positive-partial transpose will not be detected, as must be the case by construction. Nevertheless, it is important to realise conceptually the structure of the entanglement witnesses associated to negativity. In the following examples, we first calculate the negativity of two-qubit isotropic states and then show how the dual formulation of the negativity SDP can be used to detect the entanglement of all pure entangled states. That is, to prove that there are no pure PPT entangled quantum states.

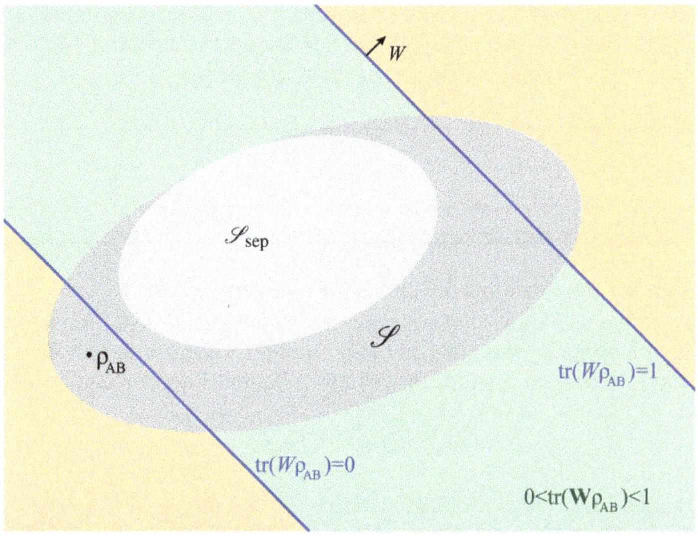

Figure 5.1. *Negativity as a quantitative entanglement witness.* Geometrically, we can view an entanglement witness W as specifying a direction in the space of quantum states; the witness is such that the set of all separable states $\mathcal{S}_{sep}$ lies in the half-space $\mathrm{tr}(W\sigma_{AB}) \geqslant 0$. Additionally, for the negativity, entanglement witnesses are normalised such that $\mathcal{S}_{sep}$ lies in the strip such that $0 \leqslant \mathrm{tr}(W\sigma_{AB}) \leqslant 1$. The value $\mathrm{tr}(W\rho_{AB})$ of an entangled quantum state ρ_{AB} can be made negative by appropriately choosing the direction of W. The negativity corresponds to (the negative of) this value, for the optimally chosen direction W.

Example 5.2 Negativity of two-qubit isotropic states.

In example 5.1 we saw how to use the PPT criterion to detect the entanglement of two-qubit isotropic states. In this example, we will use this simple state to gain some understanding of the primal and dual formulations of the negativity.

When $w > \frac{1}{3}$, in example 5.1 it was shown that $\rho_{AB}(w)^{\mathsf{T}_A}$ has a single negative eigenvalue $\lambda_- = \frac{1-3w}{4}$. The negativity of the state is therefore $\mathcal{N}(\rho_{AB}(w)) = \frac{3w-1}{4}$.

Turning now to the primal SDP formulation of negativity (5.14), a natural choice of primal variables are $Z = [\rho_{AB}(w)^{\mathsf{T}_A}]^{(+)}$ and $Y = [\rho_{AB}(w)^{\mathsf{T}_A}]^{(-)}$, the positive and negative parts of $\rho_{AB}(w)^{\mathsf{T}_A}$ respectively. In exercise 5.3 below, these are shown to be optimal choices. With these choices, the objective function evaluates to $\mathrm{tr}([\rho_{AB}(w)^{\mathsf{T}_A}]^{(-)}) = \frac{3w-1}{4}$, as expected.

Turning to the dual SDP formulation (5.15), we can take $X = |\Psi^-\rangle\langle\Psi^-|$, the projector onto the antisymmetric maximally entangled state, the eigenstate of $\rho_{AB}(w)^{\mathsf{T}_A}$ with eigenvalue -1. This is feasible, as it is both positive semidefinite, and less than the identity, being a projector. In exercise 5.3 below, this is again shown to be an optimal choice, since it is the projector onto the negative part of $\rho_{AB}(w)^{\mathsf{T}_A}$. Direct calculation shows that $\mathrm{tr}(X^{\mathsf{T}_A}\rho_{AB}(w)) = \mathrm{tr}(X\rho_{AB}(w)^{\mathsf{T}_A}) = \frac{1-3w}{4}$, as expected.

Finally, from this construction we therefore obtain the (single) quantitative entanglement witness $W = |\Psi^-\rangle\langle\Psi^-|^{T_A}$, which optimally detects the entanglement of the isotropic state for all values $w > \frac{1}{3}$, and certifies that the negativity is (at least) $\frac{3w-1}{4}$.

Example 5.3 All pure entangled states have non-zero negativity.

In this example we will show that all pure entangled states have non-zero negativity. Another way of phrasing this is to say that the PPT criterion in fact detects all pure entangled states. Consider a state $|\Psi\rangle$ written in Schmidt decomposition as

$$|\Psi\rangle = \sum_i \sqrt{p_i}|i_A\rangle|i_B\rangle, \tag{5.21}$$

where $\{|i_A\rangle\}$ and $\{|j_B\rangle\}$ are orthonormal bases for systems A and B respectively, and $\sqrt{p_i} \geqslant 0$ are the Schmidt coefficients of the state. Since the state is assumed to be entangled, at least two of the Schmidt coefficients do not vanish. Without loss of generality, we will assume that $p_1 > 0$ and $p_2 > 0$. Let us consider the potential dual variable

$$X = |\Psi_{12}^-\rangle\langle\Psi_{12}^-|, \quad |\Psi_{12}^-\rangle = \frac{1}{\sqrt{2}}(|1_A\rangle|2_B\rangle - |2_A\rangle|1_B\rangle), \tag{5.22}$$

where we see that $|\Psi_{12}^-\rangle$ is the maximally entangled singlet state in the two-qubit subspace spanned by the first and second basis states of each system.

Since X is a rank-1 projector, it follows directly that it is both a positive semidefinite operator and less than the identity, i.e. both of the constraints (5.15b) and (5.15c) are satisfied. This is thus a feasible dual variable. All that remains is to evaluate the dual objective function. To that end, we see that

$$\begin{aligned} X^{T_A} &= |\Psi_{12}^-\rangle\langle\Psi_{12}^-|^{T_A}, \\ &= \frac{1}{2}(|1_A\rangle|2_B\rangle\langle 1_A|\langle 2_B| + |2_A\rangle|1_B\rangle\langle 2_A|\langle 1_B| - |1_A\rangle|1_B\rangle\langle 2_A|\langle 2_B| - |2_A\rangle|2_B\rangle\langle 1_A|\langle 1_B|), \end{aligned}$$

and therefore a simple calculation shows that

$$-\mathrm{tr}(X^{T_A}|\Psi\rangle\langle\Psi|) = \sqrt{p_1 p_2} > 0. \tag{5.23}$$

Since the value of the dual objective function, when evaluated using a dual feasible variable places a lower bound on the value of the primal objective function—in this case the negativity of the state—this shows that all entangled pure states have non-zero negativity. In other words, this shows that all pure entangled states have a negative-partial transpose. The dual variable $X = |\Psi_{12}^-\rangle\langle\Psi_{12}^-|$ is optimal when the only non-zero Schmidt coefficients are $\sqrt{p_1}$ and $\sqrt{p_2}$. When this is not the case, i.e. when there are additional non-zero Schmidt coefficients, a sum of projectors can be used. This is both feasible and optimal, as shown in exercise 5.4.

The last example teaches us a very useful lesson: we can use the same witness to detect the entanglement of different states. This is very useful from an experimental point of view, because it allows us to detect entanglement of states that are unknown (i.e. without the need of performing quantum state tomography) by measuring a witness operator W. Furthermore, the observed violation of the witness provides a lower bound on the negativity of the state, since the witness might not be optimal for this state. In fact, this is a general idea that goes beyond the negativity: after solving an instance of an SDP for a given quantity, we can obtain the dual variables and use them to estimate the quantity for other instances.

Exercises

5.2 Show that the set $\mathcal{S}_{\text{PPT}}$ is a convex set. That is, show that if π_{AB} and π'_{AB} are both PPT states, then for all $0 \leqslant p \leqslant 1$, the state $p\pi_{AB} + (1-p)\pi'_{AB}$ is contained in $\mathcal{S}_{\text{PPT}}$.

5.3 In this exercise we will show that (5.13) constitute primal and dual SDP formulations of $\|A^{(-)}\|_1$.
- (a) Show that $Y = A^{(-)}$ and $Z = A^{(+)}$ are feasible variables for the primal SDP.
- (b) Given part (a), show that the primal optimal value α^* lower bounds $\|A^{(-)}\|_1$.
- (c) Write down the Lagrangian for the primal problem and use it to derive the dual problem.
- (d) Show that both the primal and dual problems are strictly feasible, and therefore that strong duality holds.
- (e) Show that $X = \Pi^{(-)}$ (the projector onto the negative eigenspace of A) is a feasible variable for the dual SDP.
- (f) Given part (e), show that the dual optimal value β^* upper bounds $\|A^{(-)}\|_1$.
- (g) Use parts (b), (d) and (f) to show that this pair of SDPs are primal and dual formulations of $\|A^{(-)}\|_1$.

5.4 In this exercise we will study the generalisation of (5.22), and show that it evaluates to the negativity of a pure state.
- (a) Consider the family of operators

$$X_{ij} = |\Psi_{ij}^-\rangle\langle\Psi_{ij}^-|, \quad |\Psi_{ij}^-\rangle = \frac{1}{\sqrt{2}}(|i_A\rangle|j_B\rangle - |j_A\rangle|i_B\rangle), \quad j > i. \tag{5.24}$$

Show that

$$-\text{tr}(X_{ij}^{\,\mathsf{T}_A}|\Psi\rangle\langle\Psi|) = \sqrt{p_i p_j} \tag{5.25}$$

where $|\Psi\rangle$ is the pure entangled state from (5.4).
- (b) Show that X_{ij} is orthogonal to $X_{i'j'}$ unless $i = i'$ and $j = j'$.
- (c) Consider the dual variable

$$X = \sum_{j>i} X_{ij}. \tag{5.26}$$

Using part (b), show that this dual variable is feasible for the dual negativity SDP (5.15). *Hint: It is useful to consider the operator X^2, and what it tells us about the eigenvalues of X.*
- (d) Using part (a), show that

$$-\text{tr}(X^{\mathsf{T}_A}|\Psi\rangle\langle\Psi|) = \sum_{j>i}\sqrt{p_i p_j}\,. \tag{5.27}$$

(e) It can be shown (by independent means) that the negativity of a pure state $|\Psi\rangle$ is

$$\mathcal{N}(|\Psi\rangle\langle\Psi|) = \frac{\left(\sum_i \sqrt{p_i}\right)^2 - 1}{2}. \tag{5.28}$$

Show that this expression is equal to the right-hand side of (5.27).

This shows that X is an optimal dual variable, and evaluates to the negativity of a pure state, irrespective of what the state is, i.e. irrespective of the Schmidt coefficients.

5.4 Random robustness of entanglement and SDP relaxations

We can further use the positive-partial-transpose condition to arrive at the important concept of an *SDP relaxation.* Sometimes there is no computationally efficient or easy way to solve a problem. In such instances, it is important and useful to be able to find a relaxation of the problem, which can be easily solved, and which will provide useful bounds. A common use of semidefinite programming is to find efficient relaxations of both convex and non-convex optimisation problems. This is precisely what happens with the separability problem.

The feasibility problem of testing whether or not a quantum state is separable is computationally hard in general. However, as we saw above, testing whether or not a state is PPT can be solved via an SDP. Clearly, this relaxation doesn't allow us to detect the entanglement of all states—those states which are both PPT and entangled—but we can certify the entanglement of all entangled states which are not PPT, i.e. all NPT entangled states.

It is important to realise that not only can we relax the feasibility problem, but the PPT criterion can also be used to relax optimisation problems involving entanglement too, for instance, to find bounds on entanglement quantifiers. We can illustrate this by looking at a quantifier known as the *random robustness of entanglement* $R_R(\rho_{AB})$. This is defined by the following optimisation problem

$$R_R(\rho_{AB}) = \text{minimise} \quad r \tag{5.29a}$$

$$\text{subject to} \quad \frac{\rho_{AB} + r\frac{\mathbb{1}_A \otimes \mathbb{1}_B}{d_A d_B}}{1 + r} \in \mathcal{S}_{\text{sep}}, \tag{5.29b}$$

$$r \geqslant 0. \tag{5.29c}$$

If the optimal value of this problem is 0, then the first constraint implies that $\rho_{AB} \in \mathcal{S}_{\text{sep}}$, while if it is strictly positive, then the state is necessarily entangled. The random robustness of entanglement of ρ_{AB} can be interpreted as being the minimal amount of white noise $\frac{\mathbb{1}\otimes\mathbb{1}}{d_A d_B}$ that we must mix with ρ_{AB} such that the mixture becomes a separable state.

Unfortunately, this problem cannot be solved easily, due to the fact that there is no known algorithmic characterisation of the set of separable states $\mathcal{S}_{\text{sep}}$ (except in the case of the smallest dimensions). We can, nevertheless relax this optimisation problem and replace the set $\mathcal{S}_{\text{sep}}$ by the set $\mathcal{S}_{\text{PPT}}$ of PPT states. As seen before, testing if a state is PPT can be cast as an SDP, so we can consider the following relaxed SDP:

$$R_R^{\text{NPT}}(\rho_{AB}) = \text{minimise} \quad r \tag{5.30a}$$

$$\text{subject to} \quad \left(\rho_{AB} + r\frac{\mathbb{I}_A \otimes \mathbb{I}_B}{d_A d_B}\right)^{\mathrm{T}_A} \geqslant 0, \tag{5.30b}$$

$$r \geqslant 0. \tag{5.30c}$$

Notice that in the above we have multiplied the constraint (5.29b) from previously by $1 + r$. This can be done, without any loss of generality, since $1 + r \geqslant 0$, and therefore we do not change the positive semidefiniteness of the left-hand side by performing this multiplication. Essentially what is being done here is to consider an unnormalised state, which can always be normalised later. The reason for removing the denominator should hopefully be clear—it turns the problem into an SDP, while with the denominator in place it would not yet be in the form of an SDP.

In the above we see that $R_R^{\text{NPT}}(\rho_{AB}) = r^* > 0$ means that some non-zero amount of white noise has to be mixed with ρ_{AB} to make the mixture PPT, so that ρ_{AB} is guaranteed to be entangled by the PPT criterion. If, however $r^* = 0$ then ρ_{AB} is PPT, i.e. is *not* NPT entangled. Thus, we cannot guarantee whether it is entangled or not. As such, the relaxed robustness $R_R^{\text{NPT}}(\rho_{AB})$ can be seen as the robustness to being not-NPT entangled, and can be viewed as an alternative quantifier to the entanglement negativity from section 5.3.

Crucially, in terms of the quantification of the entanglement of ρ_{AB}, since $\mathcal{S}_{\text{sep}} \subset \mathcal{S}_{\text{PPT}}$, we see that the optimal value r^* of the relaxed problem (5.30) lower bounds the original problem (5.29), i.e.

$$R_R^{\text{NPT}}(\rho_{AB}) \leqslant R_R(\rho_{AB}). \tag{5.31}$$

That is, this shows that we need to mix with less noise to make a state PPT than to make it separable. It also shows that the relaxation simply doesn't count the entanglement of PPT states. Intuitively, if the set of states $\mathcal{S}_{\text{PPT}}$ is 'similar' to the set of separable states—i.e. if the total volume of PPT entangled states is small—then this relaxation gives a reasonable approximation to the random robustness of entanglement. In practice this is a reasonable first approximation. As we will see below, better SDP relaxations do exist, which provide tighter bounds.

We end by stressing that the above is just one example of how the PPT criterion can be used to find bounds on quantities of interest. In general, in any convex optimisation problem where the constraint $\sigma_{AB} \in \mathcal{S}_{\text{sep}}$ needs to be imposed, we can always relax this constraint to the SDP constraint $\sigma_{AB} \in \mathcal{S}_{\text{PPT}}$, which in many cases leads to a relaxed problem which is itself an SDP.

Exercises

5.5 (a) Show that the random robustness of entanglement (5.29) of the two-qubit isotropic state (5.8) is

$$R_R(\rho_{AB}(w)) = 3w - 1. \tag{5.32}$$

Hint: You can compute for the random robustness analytically, without needing to perform any optimisation.

(b) Find the random robustness of the two-qudit isotropic state (5.11).

5.6 In this exercise we will derive the dual SDP formulation of (5.30) for $R_R^{\text{NPT}}(\rho_{AB})$, in order to compare with the dual formulation of the negativity from (5.17).

(a) Write down the Lagrangian associated to (5.30), by introducing dual variables X and μ for the first and second constraints respectively.

(b) Identify the constraints necessary to make the Lagrangian (i) a lower bound on the optimal primal value and (ii) independent of the primal variables.

(c) Use part (c) to arrive at the dual formulation of (5.30) and confirm that strong duality holds.

(d) Introduce a new dual variable $W = X^{\mathsf{T}_A}$, and show, after eliminating slack variables (if this has not already been done), that the dual can be expressed as

$$R_R^{\text{NPT}}(\rho_{AB}) = \text{maximise} \quad -\text{tr}(W\rho_{AB}) \tag{5.33a}$$

$$\text{subject to} \quad \text{tr}(W) \leqslant d_A d_B, \tag{5.33b}$$

$$W^{\mathsf{T}_A} \geqslant 0. \tag{5.33c}$$

(e) Explain why the constraint (5.33b) can be viewed as a normalisation constraint, demanding that value of the entanglement witness W evaluated on the maximally mixed state $\mathbb{1}/d_A d_B$ should not exceed 1, and make a sketch similar in form to figure 5.1 showing the structure of the entanglement witness geometrically.

5.7 An important variant of the robustness of entanglement is the *generalised robustness of entanglement*, given by

$$R_G(\rho_{AB}) = \text{minimise} \quad r \tag{5.34a}$$

$$\text{subject to} \quad \frac{\rho_{AB} + r\sigma_{AB}}{1 + r} \in \mathcal{S}_{\text{sep}}, \tag{5.34b}$$

$$r \geqslant 0, \tag{5.34c}$$

$$\sigma_{AB} \geqslant 0, \quad \text{tr}(\sigma_{AB}) = 1, \tag{5.34d}$$

where the 'noise' is now any (possibly entangled) quantum state σ_{AB}, and the optimisation seeks to find the worst-case noise, mixing with which makes ρ_{AB} separable as quickly as possible.

(a) Show that by relaxing the separability constraint to the PPT constraint, and by introducing the new variable $\omega_{AB} = r\sigma_{AB}$, that a relaxation of the generalised robustness of entanglement is given by the following SDP:

$$R_G^{\mathrm{NPT}}(\rho_{AB}) = \text{minimise} \quad \mathrm{tr}(\omega_{AB}) \tag{5.35a}$$

$$\text{subject to} \quad (\rho_{AB} + \omega_{AB})^{\mathsf{T}_A} \succcurlyeq 0, \tag{5.35b}$$

$$\omega_{AB} \succcurlyeq 0. \tag{5.35c}$$

(b) Write down the Lagrangian associated to (5.35), and use it to show that the dual formulation of $R_G^{\mathrm{NPT}}(\rho_{AB})$ is given by

$$R_G^{\mathrm{NPT}}(\rho_{AB}) = \text{maximise} \quad -\mathrm{tr}(W\rho_{AB}) \tag{5.36a}$$

$$\text{subject to} \quad W^{\mathsf{T}_A} \succcurlyeq 0, \tag{5.36b}$$

$$\mathbb{1} \succcurlyeq W, \tag{5.36c}$$

making sure to confirm that strong duality holds.

(c) The dual formulation (5.36) shows that the generalised robustness of NPT entanglement can be interpreted as a quantitative entanglement witness. Find the interpretation of the normalisation constraint (5.36c), and use this to make a sketch similar to figure 5.1.

5.8 In this exercise we will calculate the generalised robustness of the two-qubit isotropic state (5.8).

(a) Using the fact that $\mathbb{1} \otimes \mathbb{1} = |\Phi^+\rangle\langle\Phi^+| + |\Phi^-\rangle\langle\Phi^-| + |\Psi^+\rangle\langle\Psi^+| + |\Psi^-\rangle\langle\Psi^-|$, where $|\Phi^+\rangle$, $|\Phi^-\rangle$, $|\Psi^+\rangle$ and $|\Psi^-\rangle$ are the four two-qubit Bell states, along with knowledge that two-qubit isotropic states become separable at $w = \frac{1}{3}$, show that the generalised robustness of entanglement of $|\Phi^+\rangle$ is at most

$$R_G(|\Phi^+\rangle\langle\Phi^+|) \leqslant 1, \tag{5.37}$$

and find the noise ω_{AB}^* which achieves this value.

(b) Using part (a), show that the generalised robustness of the two-qubit isotropic state $\rho_{AB}(w)$ is at most

$$R_G(\rho_{AB}(w)) \leqslant \frac{3w-1}{2}. \tag{5.38}$$

(c) Show that the operator $W = \alpha|\Psi^-\rangle\langle\Psi^-|^{\mathsf{T}_A}$ has as eigenstates the four Bell states $|\Phi^+\Phi$, $|\Phi^-\rangle$, $|\Psi^+\rangle$ and $|\Psi^-\rangle$ with eigenvalues $-\frac{\alpha}{2}, \frac{\alpha}{2}, \frac{\alpha}{2}$ and $\frac{\alpha}{2}$ respectively. Use this to show that W is a feasible dual variable for the dual formulation of the generalised robustness of NPT entanglement (5.36) as long as $\alpha \leqslant 2$.

(d) Evaluate the dual objective function of (5.36), and show therefore that the generalised robustness of entanglement of the two-qubit isotropic state is $R_G(\rho_{AB}(w)) = \frac{3w-1}{2}$, i.e. that the bound obtained in part (b) is tight.

5.9 Another entanglement quantifier is the *weight of entanglement*. This quantifies the minimal (probabilistic) amount of entanglement that needs to be used in order to reproduce a given state. In particular, it is given by the following optimisation problem

$$W(\rho_{AB}) = \text{minimise} \quad p \tag{5.39a}$$

$$\text{subject to} \quad p\omega_{AB} + (1-p)\sigma_{AB} = \rho_{AB}, \tag{5.39b}$$

$$\omega_{AB} \succcurlyeq 0, \qquad \mathrm{tr}(\omega_{AB}) = 1, \tag{5.39c}$$

$$\sigma_{AB} \in \mathcal{S}_{\mathrm{sep}}, \tag{5.39d}$$

$$p \geqslant 0. \tag{5.39e}$$

The optimal separable state σ_{AB}^* in the decomposition (5.39b) of ρ_{AB} is referred to as the '*best-separable-approximation*' of ρ_{AB}.

(a) Show that by relaxing the separability constraint to the PPT constraint, and defining new variable $\tilde{\sigma}_{AB} = (1 - p)\sigma_{AB}$ that a relaxation of the weight of entanglement is the following SDP:

$$W^{\text{NPT}}(\rho_{AB}) = \text{minimise} \quad 1 - \text{tr}(\tilde{\sigma}_{AB}) \tag{5.40a}$$

$$\text{subject to} \quad \rho_{AB} - \tilde{\sigma}_{AB} \succcurlyeq 0, \tag{5.40b}$$

$$\tilde{\sigma}_{AB} \succcurlyeq 0, \qquad \tilde{\sigma}_{AB}^{\mathsf{T}_A} \succcurlyeq 0. \tag{5.40c}$$

(b) Write down the Lagrangian associated to (5.40), and use it to show that the dual formulation of $W^{\text{NPT}}(\rho_{AB})$ is given by

$$W^{\text{NPT}}(\rho_{AB}) = \text{maximise} \quad -\text{tr}(W\rho_{AB}) \tag{5.41a}$$

$$\text{subject to} \quad W = Y^{\mathsf{T}_A} + Z, \tag{5.41b}$$

$$Y \succcurlyeq 0, \qquad Z \succcurlyeq 0, \tag{5.41c}$$

$$W \succcurlyeq -\mathbb{1}, \tag{5.41d}$$

making sure to confirm that strong duality holds.

(c) Show that the constraints (5.41b) and (5.41c) together ensure that W is an entanglement witness, i.e. $\text{tr}(W\sigma_{AB})$ is nonnegative for all separable states σ_{AB}. *Note that this form is more general than just taking* $W^{\mathsf{T}_A} \succcurlyeq 0$, *as was the, constraint above. Such entanglement witnesses are known as 'decomposable witnesses'.*

(d) The dual formulation (5.41), along with part (c), shows that the weight of NPT entanglement can be also interpreted as a quantitative entanglement witness. Find the interpretation of the normalisation constraint (5.41d), and use this to make a sketch similar to figure 5.1.

5.5 *k*-symmetric extensions

We have seen above that the PPT criterion gives a relaxation of the separability constraint in terms of a semidefinite constraint. This allowed us to obtain lower bounds on quantifiers of entanglement, such as the random robustness of entanglement, via SDPs. A natural question is whether there are better relaxations of the separability constraint that allow us to obtain tighter bounds on entanglement quantifiers, but which nevertheless remain SDPs. In this section we will see that this is indeed the case, and that there is a sequence of relaxations—each one tighter than the previous—which provide bounds on entanglement quantifiers. This sequence of relaxations is captured by the notion of a *k-symmetric extension* of a quantum state. Such extensions provides a hierarchy of SDPs that constitutes better and better lower bounds on the entanglement of quantum states.

We say that a bipartite state ρ_{AB} has a k-symmetric extension if there exists a $(k+1)$-partite state $\rho_{AB_1\ldots B_k}$ such that all of the two-party reduced density operators ρ_{AB_j} coincide, and are equal to the state ρ_{AB}, that is if $\rho_{AB_j} = \rho_{AB}$ for all j. An extremely important result is the following:

> A quantum state ρ_{AB} is separable if and only if it has a k-symmetric extension for all k.

The 'if' statement can be proven easily by construction. Consider an arbitrary separable state,

$$\sigma_{AB}^{\text{sep}} = \sum_{\lambda} p(\lambda)\rho_\lambda^A \otimes \rho_\lambda^B. \tag{5.42}$$

A k-symmetric extension of this state can directly be written down, by simply 'copying' the state of the B system k times. That is, the state

$$\rho_{AB_1\ldots B_k} = \sum_{\lambda} p_\lambda \rho_\lambda^A \otimes \rho_\lambda^{B_1} \otimes \cdots \otimes \rho_\lambda^{B_k}, \tag{5.43}$$

with $\rho_\lambda^{B_j} = \rho_\lambda^B$ for all values of j, has the property that $\rho_{AB_j} = \sigma_{AB}^{\text{sep}}$. The intuition behind this is that since the two systems are only classically correlated (through the variable λ and their local states), it is possible to copy or 'clone' these correlations and produce a k-symmetric extension.

In the other direction, the proof—that *only* separable states have k-symmetric extensions for all values of k—is not as straightforward. This property nevertheless captures—and quantifies—the notion of *monogamy of entanglement* that we encountered in section 3.2 in the context of the quantum marginal problem. Monogamy of entanglement is the phenomenon whereby if ρ_{AB} is entangled, then this limits the amount of entanglement that the A can have with other systems at the same time. That is, A only has a finite capacity of sharing entanglement. So, sharing a significant amount of entanglement with one system limits the amount of entanglement that can be shared with other systems (this is in stark contrast to classical correlations, which have no such finite capacity).

Thus, since a k-symmetric extension requires A to share exactly the same amount of entanglement with each B, monogamy of entanglement shows us that for some k we must exceed the capacity of A to be entangled, and such an extension cannot exist. Moreover, if A is highly entangled with B, then it should fail to have k-symmetric extensions for small values of k compared to less entangled states, as the capacity will be reached much faster.

This is the key insight behind k-symmetric extensions, and shows how they can be used as a powerful tool for entanglement detection. In particular, if we flip the above logic, and consider the set of states that have a k-symmetric extension, then as k increases states in this set cannot have much entanglement, so they must be approximately separable. As k increases, the entanglement drops, and the approximation becomes better.

Before proceeding further, it will be useful to focus on the one aspect of k-symmetric extensions that hasn't been discussed yet—the *symmetry* in this definition. In fact we haven't yet imposed any symmetry on the extension $\rho_{AB_1\cdots B_k}$, other than demanding that all of the bipartite reduced density operators coincide. It is useful to demand that the extension is *symmetric* among all of the B subsystems. That is, if we denote by $\Pi_{\rm sym}$ the projector onto the *symmetric subspace*[2], then we demand that

$$(\mathbb{1}_A \otimes \Pi_{\rm sym})\rho_{AB_1\cdots B_k}(\mathbb{1}_A \otimes \Pi_{\rm sym}) = \rho_{AB_1\cdots B_k}. \tag{5.44}$$

This says that the state is symmetric under permutation of any of the B subsystems, which is much stronger than just demanding that each of the reduced density operators ρ_{AB_j} coincide. There are a couple of reasons for imposing this symmetry constraint. First, at a conceptual level, it is important to realise that this is all that is needed from the extension; it isn't necessary to consider non-symmetric states[3]. Second, and probably more importantly, this makes k-symmetric extensions a much more efficient tool from the perspective of SDPs (which we will get to shortly). In particular, if the dimension of Alice's and Bob's systems are taken to be the same, and equal to d, then the naive dimension of a k-symmetric extension is d^{k+1}, which grows exponentially fast with k. Very quickly such extensions become too big to use in any actual piece of code. It is important therefore to understand to what extent this can be overcome, and this is where the symmetry comes in. By restricting to the symmetric subspace, we do not need to consider a state in the full, exponentially-large Hilbert space, but rather a state in the symmetric subspace, which is of much smaller dimension, $d\binom{d+k-1}{k}$. This dimension only grows polynomially with k, as k^{d-1} when k is large, and so it is much more efficient to use this method when taking into account the symmetry.

We can now put everything together, and define the set of states which have a k-symmetric extension (similar to how the set of PPT states was defined in (5.6)),

$$\begin{aligned}\mathcal{S}_{k\rm SE} = \Big\{\sigma_{AB} \mid \sigma_{AB} = \rho_{AB_1}, \rho_{AB_1\cdots B_k} \succcurlyeq 0,\\ (\mathbb{1}_A \otimes \Pi_{\rm sym})\rho_{AB_1\cdots B_k}(\mathbb{1}_A \otimes \Pi_{\rm sym}) = \rho_{AB_1\cdots B_k}\Big\}.\end{aligned} \tag{5.45}$$

Note that we only need to impose $\sigma_{AB} = \rho_{AB_1}$, since due to the symmetry of the state, it then follows immediately that $\sigma_{AB} = \rho_{AB_j}$ for all j. Note also that $\rho_{AB_1\cdots B_k}$ satisfies $\mathrm{tr}(\rho_{AB_1\cdots B_k}) = 1$, since $\sigma_{AB} = \rho_{AB_1}$. We refer to the set of states which have a k-symmetric extension as being *k-extendible*.

[2] There are many ways of representing the projector onto the symmetric subspace. Two useful forms are

(i) $\Pi_{\rm sym} = \int d\psi |\psi\rangle\langle\psi|^{\otimes k}$, where $d\psi$ is the invariant (Haar) measure over the state space;

(ii) $\Pi_{\rm sym} = \frac{1}{k!}\sum_{\pi\in S_k} V_\pi$, where S_k is the symmetric group over k elements, and V_π is the permutation unitary such that $V_\pi|i_1, \ldots, i_k\rangle = |i_{\pi^{-1}(1)}, \ldots, i_{\pi^{-1}(k)}\rangle$.

[3] One way to see this is to note that if we find an asymmetric extension, we can always use it to construct other asymmetric extensions by performing permutations of the B systems. These can then all be mixed equally, which produces a symmetric extension, which will be just as good as any of the individual asymmetric ones. We can therefore restrict to symmetric extensions from the beginning without loss of generality.

Crucially for our purposes here, the set of k-extendible states $\mathcal{S}_{k\mathrm{SE}}$ is specified by a set of positive semidefinite constraints and linear matrix equalities. As such, this can be specified inside an SDP to define a feasible set. This allows us to use the k-extendible set as an approximation to the set of separable states, which then allows us to relax non-SDP optimisation problems involving the set of separable states to SDPs involving k-extendible states.

In contrast to before, when we considered the PPT-relaxation (5.6) of the separable set (5.7), instead of getting a single approximation, we now obtain a sequence of relaxations, indexed by k. As k increases, the set of k-extendible states becomes a better approximation of the separable set, and in the limit $k \to \infty$ the sets coincide. This thus provides us with an extremely powerful approximation scheme, which is highly useful in practice.

The problem of whether a state ρ_{AB} has a k-symmetric extension is an instance of a feasibility semidefinite program. In particular, it is the following problem:

$$\text{find} \quad \rho_{AB_1\cdots B_k} \tag{5.46a}$$

$$\text{subject to} \quad \rho_{AB} = \rho_{AB_1}, \tag{5.46b}$$

$$\rho_{AB_1\cdots B_k} \geqslant 0, \tag{5.46c}$$

$$(\mathbb{1}_A \otimes \Pi_{\mathrm{sym}})\rho_{AB_1\cdots B_k}(\mathbb{1}_A \otimes \Pi_{\mathrm{sym}}) = \rho_{AB_1\cdots B_k}. \tag{5.46d}$$

In contrast to the feasibility SDP for being PPT that we saw in (5.10), which was purely academic and not useful in practice, here this feasibility SDP is necessary and useful for determining when a state has a k-symmetric extension. Note also that in the above SDP we have written the constraint that $\rho_{AB_1\cdots B_k}$ is in the symmetric subspace explicitly in (5.46d). In practice (i.e. when writing code), it is crucial to work *directly* in the symmetric subspace (and not impose it as a constraint), in order to benefit from the size reduction that this offers in the SDP. It is nevertheless useful to write the SDP in the form given above, to conceptually emphasise the symmetric nature of the extension.

As with the PPT criterion, the k-symmetric criterion is most powerful when used for approximately evaluating properties of a state, such as entanglement quantifiers. As a concrete example, let us return to the random robustness of entanglement (5.29). We can now consider the following sequence of relaxations

$$R_R^{k\mathrm{SE}}(\rho_{AB}) = \text{minimise} \quad r \tag{5.47a}$$

$$\text{subject to} \quad \frac{\rho_{AB} + r\dfrac{\mathbb{1}_A \otimes \mathbb{1}_B}{d_A d_B}}{1 + r} \in \mathcal{S}_{k\mathrm{SE}}, \tag{5.47b}$$

$$r \geqslant 0. \tag{5.47c}$$

As is shown in exercise 5.11, this can be re-expressed as the following SDP:

$$R_R^{k\mathrm{SE}}(\rho_{AB}) = \text{minimise} \quad r \tag{5.48a}$$

$$\text{subject to} \quad \rho_{AB} + r\frac{\mathbb{1}_A \otimes \mathbb{1}_B}{d_A d_B} = \tilde{\rho}_{AB_1}, \tag{5.48b}$$

$$\tilde{\rho}_{AB_1\cdots B_k} \succcurlyeq 0, \tag{5.48c}$$

$$(\mathbb{1}_A \otimes \Pi_{\mathrm{sym}})\tilde{\rho}_{AB_1\cdots B_k}(\mathbb{1}_A \otimes \Pi_{\mathrm{sym}}) = \tilde{\rho}_{AB_1\cdots B_k}, \tag{5.48d}$$

$$r \geqslant 0. \tag{5.48e}$$

This sequence of SDPs for increasing k provides us with a sequence of lower-bounds on the random robustness of entanglement, which converge in the limit,

$$R_R^{1\mathrm{SE}}(\rho_{AB}) \leqslant R_R^{2\mathrm{SE}}(\rho_{AB}) \leqslant \cdots \leqslant \lim_{k\to\infty} R_R^{k\mathrm{SE}}(\rho_{AB}) = R_R(\rho_{AB}). \tag{5.49}$$

In practice, since the SDPs become larger in size (optimising over extensions of larger dimension as k increases), there is a trade-off between the resources required (the time to solve the SDP or the memory needed in order to solve it) and the level of approximation achieved.

Exercises

5.10 In this exercise we will show that two-qubit isotropic states (5.8) have 2-symmetric extensions when $w \leqslant \frac{1}{2}$.

(a) Consider the following tripartite state,

$$\rho_{AB_1B_2} = \frac{1}{2}|\Phi^+\rangle\langle\Phi^+|_{AB_1} \otimes \frac{\mathbb{1}_{B_2}}{2} + \frac{1}{2}|\Phi^+\rangle\langle\Phi^+|_{AB_2} \otimes \frac{\mathbb{1}_{B_1}}{2}, \tag{5.50}$$

where we note that on the right-hand side, since the systems are labelled, we do not attribute any significance to the order of the tensor factors. Verify that this state is symmetric among the B systems, and that it is a 2-extension for the isotropic state $\rho_{AB}(\frac{1}{2})$.

(b) Show that by mixing $\rho_{AB_1B_2}$ with the maximally mixed state of three qubits, that we can furthermore obtain a 2-extension for any isotropic state with $w \leqslant \frac{1}{2}$.

5.11 In this exercise we will show that (5.47) can be cast as the SDP (5.48).

(a) Using the definition of the set $\mathcal{S}_{k\mathrm{SE}}$, write down the optimisation problem that needs to be solved in order to calculate the random robustness of being k-extendible. *Note that this will not yet be an SDP. Why not?*

(b) Introduce a new variable $\tilde{\rho}_{AB_1\cdots B_k} = (1 + r)\rho_{AB_1\cdots B_k}$, and show that in terms of this variable, the optimisation problem from part (a) becomes equal to (5.48).

We can interpret the SDP (5.48) as optimising over the set of super-normalised k-symmetric extensions.

We can notice in the above that the first approximation, $R_R^{1\mathrm{SE}}(\rho_{AB})$ is in fact always equal to 0. This is because a 1-extension is equivalent to not seeking an extension at all, and in this case the set $\mathcal{S}_{1\mathrm{SE}}$ is equal to the set of all bipartite states, and does not distinguish between separable and entangled states at all. We can improve this, by combining the idea of a k-symmetric extension with the PPT criterion from section 5.2.

In particular, beyond just demanding that the extension is symmetric, we can furthermore impose that it should be positive under partial transposition. Up until now we have only considered the PPT criterion for bipartite systems, whereas we now have a multipartite system. Looking back at the symmetric extension of a separable state in (5.43), it can be noticed that the extension is a fully separable state, with no entanglement whatsoever. It follows that if we consider an arbitrary bipartition of the state into two parts, and consider taking the partial transpose of one half, the density operator will remain positive semidefinite—this in effect treats the multipartite state as a bipartite state, and applies the logic of the PPT criterion. Another way of saying the above, is that if any subset of B systems is picked, and a partial transpose is performed, then a positive semidefinite operator is obtained. We can impose this as an additional set of constraints which must be satisfied by an extension—that it is PPT when any subset of B systems is transposed. This leads to a better relaxation of separability, since the set of permissible extensions is reduced to be smaller in size.

There is however an amount of redundancy which should first be removed, which has arisen once again from the symmetry of the extension. The symmetry implies that the PPT constraints only need to be imposed for a specific subset of bipartitions, rather than for all bipartitions. To see why, consider imposing that the extension is PPT when transposing only B_1, $\rho_{AB_1\cdots B_k}^{\mathrm{T}_{B_1}} \succcurlyeq 0$. Now, because of the symmetry, we can relabel which B system is which. Consider therefore relabelling the first and second subsystems of B, $B_1 \leftrightarrow B_2$. After relabelling, it is now imposed that the partial transpose on B_2 leaves the density operator positive semidefinite. That is, because of the symmetry, we see that transposing B_1 is equivalent to transposing B_2, or *any individual B subsystem.* By a similar logic, transposing the first two subsystems is equivalent to transposing two arbitrary subsystems, and so forth. Therefore, without any loss of generality, if it is demanded that the state remains PPT after transposing the first ℓ subsystems, this is equivalent to demanding the state is PPT after transposing an arbitrary set of ℓ subsystems; we do not need to independently impose these constraints.

It is important to note that the above observation dramatically reduces the number of constraints that need to be imposed, and therefore dramatically reduces the *number of variables in the dual problem.* In particular, if we didn't take into account the symmetries, the number of bipartitions that would need to be considered is $2^{k-1} - 1$, and grows exponentially with the size of the extension. Taking into account the symmetries, only $k - 1$ bipartitions need to be considered.

Putting everything together, we can now define a final sequence of sets which approximate the set of separable states (5.7), the set of states which have a PPT k-symmetric extension, or *PPT k-extendible* states for short,

$$\begin{aligned}\mathcal{S}_{k\mathrm{SE}}^{\mathrm{PPT}} = \Big\{\sigma_{AB} \mid \sigma_{AB} = \rho_{AB_1}, \rho_{AB_1\cdots B_k} \succcurlyeq 0, (\mathbb{1}_A \otimes \Pi_{\mathrm{sym}})\rho_{AB_1\cdots B_k}(\mathbb{1}_A \otimes \Pi_{\mathrm{sym}}) \\ = \rho_{AB_1\cdots B_k}, \rho_{AB_1\cdots B_k}^{\mathsf{T}_{(\ell)}} \succcurlyeq 0 \text{ for } \ell = 1, \dots, k\Big\},\end{aligned} \tag{5.51}$$

where $\mathsf{T}_{(\ell)} = \mathsf{T}_{B_1\cdots B_\ell}$ is shorthand for transposing the first ℓ subsystems of B.

Similarly to above, the problem of determining whether a state has a PPT k-symmetric extension is the following feasibility SDP:

$$\text{find} \quad \rho_{AB_1\cdots B_k} \tag{5.52a}$$

$$\text{subject to} \quad \rho_{AB} = \rho_{AB_1}, \tag{5.52b}$$

$$\rho_{AB_1\cdots B_k} \succcurlyeq 0, \tag{5.52c}$$

$$(\mathbb{1}_A \otimes \Pi_{\mathrm{sym}})\rho_{AB_1\cdots B_k}(\mathbb{1}_A \otimes \Pi_{\mathrm{sym}}) = \rho_{AB_1\cdots B_k}, \tag{5.52d}$$

$$\rho_{AB_1\cdots B_k}^{\mathsf{T}_{(\ell)}} \succcurlyeq 0 \qquad \ell = 1, \dots, k. \tag{5.52e}$$

This can be seen as an adaptation of the feasibility SDP (5.46) for k-extendibility, adding in the PPT constraints (5.52e). In a completely analogous fashion, by adding the same constraints to (5.48), we obtain an SDP relaxation, computing the random robustness of a state to having a PPT k-symmetric extension. This same idea can be applied in many contexts, and provides a converging sequence of approximations for many quantities of interest.

5.6 Concluding remarks

In this chapter we have discussed entanglement from the perspective of SDP. As was seen, even though detecting or quantifying the entanglement of states of arbitrary dimension is a difficult task, we can nevertheless use SDPs to obtain sufficient criteria for entanglement detection, and also to obtain bounds for various entanglement quantifiers. In this regards, in this chapter we have learnt the following:

- **Entanglement negativity.** The entanglement negativity can be written as an SDP (5.14).
- **Entanglement witnesses.** Building on an idea already presented at section 3.1.5, we have seen how the dual formulation of entanglement quantifiers, such as the negativity or robustnesses, provide *quantitative* entanglement witness—see equation (5.17).
- **SDP relaxation.** Although deciding whether a state is entangled or not is a difficult task in general, we can relax the problem, in order to find an SDP formulation that approximates it. We have applied this idea to the *random robustness of entanglement*, where the separability constraint have been substituted by the PPT constraint—see equation (5.30).
- **k-symmetric extensions.** We have also presented a sequence of SDP relaxations to the separability problem that detects the entanglement of any state in the limit of large k. Although this limit is not practical, since the size of the

corresponding SDP grows with k, it nevertheless demonstrates an important point—that a hard problem can be approximated by a sequence of SDPs.

5.7 Further reading

- Horodecki R, Horodecki P, Horodecki M and Horodecki K 2009 Quantum entanglement *Rev. Mod. Phys.* **81** 865
- Gühne O and Tóth G 2009 Entanglement detection *Phys. Rep.* **474** 1
- Doherty A C 2014 Entanglement and the shareability of quantum states *J. Phys A: Math. Theor.* **47** 424004

5.8 Advanced topics

5.8.1 Entanglement witnesses from *k*-symmetric extensions

Here we briefly discuss how duality can be used in order to obtain a converging sequence of sets of entanglement witnesses, which can also prove useful in practice.

Recall that an entanglement witness is an operator which has a nonnegative expectation value on any separable state, but can have a negative expectation value for some entangled states. Formally, we can define the *set of all entanglement witnesses* as the following set

$$\mathcal{W}_{\text{ent}} = \left\{ W \mid \operatorname{tr}(W\sigma_{AB}) \geqslant 0 \text{ for all } \sigma_{AB} \in \mathcal{S}_{\text{sep}} \right\}. \tag{5.53}$$

The difficulty in characterising the set of separable states $\mathcal{S}_{\text{sep}}$ carries across into the realm of entanglement witnesses, and it is just as difficult to characterise the set of entanglement witnesses. Section 5.5 however provides us with a method for constructing *approximations* to the set of entanglement witnesses. In particular, we will see that we can obtain *inner approximations* to the set of entanglement witnesses—i.e. subsets of the set $\mathcal{W}_{\text{ent}}$, which can witness some—but not all—entangled states.

The simplest, and smallest, inner approximation to $\mathcal{W}_{\text{ent}}$ that we have already encountered implicitly in section 5.3 is the set of NPT-entanglement witnesses, arising from the PPT approximation $\mathcal{S}_{\text{PPT}}$. In particular, a summary of what was found there is that

$$\mathcal{W}_{\text{NPT}} = \left\{ W \mid \operatorname{tr}(W\sigma_{AB}) \geqslant 0 \text{ for all } \sigma_{AB} \in \mathcal{S}_{\text{PPT}} \right\}, \tag{5.54a}$$

$$= \left\{ W \mid W^{\text{T}_A} \geqslant 0 \right\}, \tag{5.54b}$$

where the first line is the definition of an NPT-entanglement witness, and the second line is the result that this set has a simple characterisation, as the set of operators which are themselves PPT. In particular, we can take $W = |\phi\rangle\langle\phi|^{\text{T}_A}$, i.e. to be the partial transpose of a rank-1 projector onto an entangled state. On the one hand, such an operator will have a nonnegative expectation value on all PPT states. On the other hand, it will detect any NPT entangled state such that $|\phi\rangle$ is contained in the eigenspace associated to the negative eigenvalue(s) of the partially transposed state.

We can obtain a sequence of better approximations to the set of entanglement witnesses by considering the set $\mathcal{S}_{kSE}$ of k-extendible states. Although it is possible to work directly with the feasibility SDP (5.46) to do this, it is preferable to work with a standard optimisation SDP when using the Lagrangian to pass to the dual. We will therefore derive the dual formulation of the k-symmetric approximation of random robustness of entanglement (5.48), and see how we can use this to find the associated approximate set of entanglement witnesses. The Lagrangian associated to (5.48) is

$$\begin{aligned}\mathcal{L} = r - \operatorname{tr}\left[W\left(\rho_{AB} + r\frac{\mathbb{I}_A \otimes \mathbb{I}_B}{d_A d_B} - \tilde{\rho}_{AB_1}\right)\right] - \operatorname{tr}(X\tilde{\rho}_{AB_1\cdots B_k}) \\ + \operatorname{tr}\left\{Y\left[(\mathbb{I}_A \otimes \Pi_{\text{sym}})\tilde{\rho}_{AB_1\cdots B_k}(\mathbb{I}_A \otimes \Pi_{\text{sym}}) - \tilde{\rho}_{AB_1\cdots B_k}\right]\right\} - xr,\end{aligned} \tag{5.55a}$$

$$\begin{aligned}= \operatorname{tr}\left\{\tilde{\rho}_{AB_1\cdots B_k}\left[W \otimes \mathbb{I}_{B_2\cdots B_k} - X + (\mathbb{I}_A \otimes \Pi_{\text{sym}})Y(\mathbb{I}_A \otimes \Pi_{\text{sym}}) - Y\right]\right\} \\ + r\left[1 - \frac{\operatorname{tr}(W)}{d_A d_B} - x\right] - \operatorname{tr}(W\rho_{AB}),\end{aligned} \tag{5.55b}$$

where $X \succcurlyeq 0$ and $x \geqslant 0$, as these are the dual variables associated with the inequality constraints. From the Lagrangian, we can see that only W will appear in the dual objective function, and hence X and x are both slack variables. After removing them from the problem, we arrive at the following initial form of the dual problem

$$\text{maximise} \quad -\operatorname{tr}(W\rho_{AB}) \tag{5.56a}$$

$$\text{subject to} \quad \frac{\operatorname{tr}(W)}{d_A d_B} \leqslant 1 \tag{5.56b}$$

$$W \otimes \mathbb{I}_{B_2\cdots B_k} + (\mathbb{I}_A \otimes \Pi_{\text{sym}})Y(\mathbb{I}_A \otimes \Pi_{\text{sym}}) - Y \succcurlyeq 0. \tag{5.56c}$$

An important simplification that can be made is to transform the final constraint (5.56c) into a more useful and insightful form. To do so, we can multiply the constraint, from the left and from the right by $\mathbb{I}_A \otimes \Pi_{\text{sym}}$. Since $\mathbb{I}_A \otimes \Pi_{\text{sym}}$ is a projector, it follows that $(\mathbb{I}_A \otimes \Pi_{\text{sym}})(\mathbb{I}_A \otimes \Pi_{\text{sym}}) = \mathbb{I}_A \otimes \Pi_{\text{sym}}$, and hence we arrive at

$$(\mathbb{I}_A \otimes \Pi_{\text{sym}})W \otimes \mathbb{I}_{B_2\cdots B_k}(\mathbb{I}_A \otimes \Pi_{\text{sym}}) \succcurlyeq 0, \tag{5.57}$$

where we use the fact that conjugating a positive semidefinite operator (multiplying on the left and right by another operator) leaves it positive semidefinite, as shown in exercise 5.12. This form is simpler than (5.56c), and helps to demonstrate its significance; it says that although the operator $W \otimes \mathbb{I}_{B_2\cdots B_k}$ need not be positive semidefinite—and in fact cannot be positive semidefinite if it will detect the entanglement of some state—its projection onto the symmetric subspace of B_1 to B_k must be a positive semidefinite operator.

Putting everything together, we thus arrive at the simplified dual SDP

$$\text{maximise } -\mathrm{tr}(W\rho_{AB}) \tag{5.58a}$$

$$\text{subject to } \mathrm{tr}(W) \leqslant d_A d_B, \tag{5.58b}$$

$$(\mathbb{1}_A \otimes \Pi_{\mathrm{sym}})W \otimes \mathbb{1}_{B_2\cdots B_k}(\mathbb{1}_A \otimes \Pi_{\mathrm{sym}}) \succcurlyeq 0, \tag{5.58c}$$

where in (5.58b) we have multiplied both sides by $d_A d_B$ to simplify. The final realisation that can be made is that here exactly the same constraint (5.62b) applies as in the dual formulation of the random robustness of NPT entanglement (5.33b). This is a normalisation constraint on the entanglement witness, and is what turns it into a *quantitative witness*, here such that the value of the witness equals the random robustness of being k-extendible. It is thus the *remaining* constraint which captures the notion of an entanglement witness for k-extendibility. To confirm this, we can consider multiplying the constraint (5.58c) by the extension $\sigma_{AB_1\cdots B_k}$ of an arbitrary k-extendible state σ_{AB_1}, and taking the trace, to obtain

$$\begin{aligned}\mathrm{tr}\left[(\mathbb{1}_A \otimes \Pi_{\mathrm{sym}})W \otimes \mathbb{1}_{B_2\cdots B_k}(\mathbb{1}_A \otimes \Pi_{\mathrm{sym}})\sigma_{AB_1\cdots B_k}\right] &= \mathrm{tr}[(W \otimes \mathbb{1}_{B_2\cdots B_k})\sigma_{AB_1\cdots B_k}] \\ &= \mathrm{tr}(W\sigma_{AB_1}) \geqslant 0,\end{aligned} \tag{5.59}$$

where in the first line the cyclicity of the trace has been used, and the symmetry of the extension, $(\mathbb{1}_A \otimes \Pi_{\mathrm{sym}})\sigma_{AB_1\cdots B_k}(\mathbb{1}_A \otimes \Pi_{\mathrm{sym}}) = \sigma_{AB_1\cdots B_k}$, and the final inequality holds as we took the trace of the product of two positive semidefinite operators. This shows that $\mathrm{tr}(W\sigma_{AB})$ is nonnegative for any k-extendible state, as required for a witness of entanglement based upon k-extendibility. We thus arrive at the following sequence of inner approximations to $\mathcal{W}_{\mathrm{ent}}$,

$$\mathcal{W}_{k\mathrm{SE}} = \left\{W \mid (\mathbb{1}_A \otimes \Pi_{\mathrm{sym}})W \otimes \mathbb{1}_{B_2\cdots B_k}(\mathbb{1}_A \otimes \Pi_{\mathrm{sym}}) \succcurlyeq 0\right\}. \tag{5.60}$$

Finally, in exercise 5.13, a similar analysis is carried out, based upon k-symmetric PPT extensions, where it is shown that the corresponding sequence of inner approximations to the set of entanglement witnesses is

$$\begin{aligned}\mathcal{W}_{k\mathrm{SE}}^{\mathrm{PPT}} = \Bigg\{ & W \mid (\mathbb{1}_A \otimes \Pi_{\mathrm{sym}})W \otimes \mathbb{1}_{B_2\cdots B_k}(\mathbb{1}_A \otimes \Pi_{\mathrm{sym}}) \succcurlyeq (\mathbb{1}_A \otimes \Pi_{\mathrm{sym}}) \\ & \sum_{\ell=1}^{k-1} X_\ell^{\mathsf{T}_\ell}(\mathbb{1}_A \otimes \Pi_{\mathrm{sym}}), \quad X_\ell^{\mathsf{T}_\ell} \succcurlyeq 0 \text{ for } \ell = 1, \dots, k-1 \Bigg\}.\end{aligned} \tag{5.61}$$

The difference to the set of witnesses for non-k-extendibility is that now $W \otimes \mathbb{1}_{B_1\cdots B_k}$ no longer needs to be positive semidefinite when projected onto the symmetric subspace of B, but must be 'greater than' the operator $\sum_{\ell=1}^{k-1} X_\ell^{\mathsf{T}_\ell}$ (on the subspace), which in general will not be a positive semidefinite operator. This is thus a relaxation of the constraint, and hence the set of witnesses, for a given k, will be *larger*, i.e. this is a better inner approximation. This is as it should be, as the corresponding approximation to the set of separable states $\mathcal{S}_{k\mathrm{SE}}^{\mathrm{PPT}}$ is a better outer approximation to the set of separable states compared to $\mathcal{S}_{k\mathrm{SE}}$. The price that we pay is that this set is computationally more demanding, with $k - 1$ additional bipartite operator variables introduced.

Exercises

5.12 (a) Show that if an operator A is positive semidefinite, $A \succcurlyeq 0$, then the conjugated operator $A' = B^{\dagger}AB$ (for an arbitrary, not necessarily Hermitian operator B) is also positive semidefinite.

Hint: It is useful to use the fact that an operator is positive semidefinite if it has a nonnegative expectation value for all states, i.e. $A \succcurlyeq 0$ is equivalent to $\langle\psi|A|\psi\rangle \succcurlyeq 0$ for all $|\psi\rangle$.

(b) Use part (a) to explain why (5.61) holds.

5.13 In this exercise we will derive the set of entanglement witnesses that detect non-PPT k-extendibility given in (5.65).

(a) Write down the SDP relaxation of the random robustness of entanglement (5.29) based upon PPT k-symmetric extensions.

(b) Write down the Lagrangian associated to this SDP, in analogy to (5.55).

(c) Use the Lagrangian from part (b) to arrive at a dual SDP formulation of the primal SDP from part (a).

(d) Simplify the dual, in analogy to how the dual (5.56) was simplified to arrive at (5.62).

(e) Use the simplified dual from part (d) to identify the set $\mathcal{W}_{k\mathrm{SE}}^{\mathrm{PPT}}$ of witnesses of non-PPT k-extendibility.

IOP Publishing

Paul Skrzypczyk and Daniel Cavalcanti

Chapter 6

Measurement incompatibility

In this chapter we return to the study of measurements, but this time rather than considering a single measurement at a time, we will consider scenarios involving *sets* of measurements, and the problem of *measurement incompatibility*—one of the fundamental aspects of quantum theory.

Conceptually, the key idea is whether a set of measurements can (or can't) be *performed simultaneously*. In other words, if a single experiment can be carried out which allows for the results of a set of measurements to be inferred simultaneously or not. Originally, measurement incompatibility was associated with the fact that some measurements are described by observables that don't commute. In fact, for projective measurements two observables commute if and only if they share a common eigenbasis. Then by performing the single experiment that implements a measurement in this eigenbasis one can obtain information about all observables that share this eigenbasis. However, when measurements are not projective (i.e. when they are described by more general Positive Operator–Valued Measure (POVM) elements), then this construction is no longer enough and we have to consider a more general notion of incompatibility, known as non-joint measurability.

In this chapter we will see that the set of all compatible sets of measurements—also referred to as *jointly measurable* sets—can be represented by a semidefinite program (SDP). We will see that this allows us to use SDP to *quantify* how incompatible a set of measurements are. In the special case of two dichotomise (two-outcome) measurement, in the 'Advanced topics' section we will use SDP duality to furthermore present a surprising result—that the measurements are incompatible precisely when they can be used to demonstrate quantum nonlocality.

6.1 Joint measurability as an SDP

Consider a set of m measurements, $\{\mathbb{M}_x\}_x$, each of which is specified by a collection of o POVM elements, $\mathbb{M}_x = \{M_{a|x}\}$ for $a = 1, \ldots, o$, where for simplicity it is

doi:10.1088/978-0-7503-3343-6ch6 © IOP Publishing Ltd 2023

assumed that each measurement has the same number of outcomes[1]. This set of measurements is *compatible* or *jointly measurable* if there exists a single *parent* measurement which can be measured in place of the individual measurements, the outcome of which is then used to simulate the outcomes of the individual measurements. More concretely, this means that there is a parent measurement $\mathbb{N} = \{N_\lambda\}$, with an *arbitrary* number of outcomes l, and a probabilistic assignment of outcomes for the m measurements, specified by a conditional probability distribution $p(a|\lambda, x)$. Thus, mathematically the set $\{\mathbb{M}_x\}_x$ is jointly measurable if every POVM element can be decomposed as

$$M_{a|x} = \sum_\lambda p(a|\lambda, x)N_\lambda. \tag{6.1}$$

A basic question that arises is: given a set of measurements $\mathbb{M}_x = \{M_{a|x}\}$, how can we determine if it is compatible. This question can be phrased as the follow feasibility problem

$$\text{find} \quad \mathbb{N}, p(a|\lambda, x) \tag{6.2a}$$

$$\text{subject to} \quad M_{a|x} = \sum_\lambda p(a|\lambda, x)N_\lambda \qquad \forall a, x, \tag{6.2b}$$

$$N_\lambda \succcurlyeq 0 \quad \forall \lambda, \qquad \sum_\lambda N_\lambda = \mathbb{1}, \tag{6.2c}$$

$$p(a|\lambda, x) \geqslant 0 \quad \forall a, x, \lambda, \qquad \sum_a p(a|\lambda, x) = 1 \quad \forall \lambda, x. \tag{6.2d}$$

This feasibility problem is not an SDP, since the constraint (6.2b) is non-linear in the variables of the problem. However, as with all the other problems encountered thus far, we can overcome this problem, and show that checking whether a set of measurements is jointly measurable or not can indeed be cast as a feasibility SDP.

The key observation to make is that we can restrict to a special class of parent measurements $\mathbb{N}$ and assignments $p(a|\lambda, x)$ without loss of generality. We can understand (6.1) operationally, interpreting it as a model that first measures $\mathbb{N}$ producing a measurement result λ, which is then used to generate the results of any measurement $\mathbb{M}_x$ probabilistically according to $p(a|\lambda, x)$. Let us think about the collection of all m measurement results jointly. When a set of measurements is jointly measurable, in principle all of the measurement results could be asked for, rather than just a single result. In order to distinguish them, we can label the outcome of the first measurement by a_1, the outcome of the second by a_2, and so forth, and collect all m into a vector $\vec{a} = (a_1, \dots, a_m)$. How many different sets of

[1] Notice that we can always assume that the measurements have the same number of outcomes without loss of generality, since, if they don't, we can simply redefine them by adding null measurement operators (corresponding to outcomes that never occur) until the number of outcomes coincide.

results could in principle be obtained? There are precisely o^m results—each measurement having o outcomes, and there being m measurements. The probability for a particular string $\vec{a}$ being produced (given the result of the measurement λ) is

$$p(\vec{a}|\lambda) = p(a_1|\lambda, x = 1)p(a_2|\lambda, x = 2)\cdots p(a_m|\lambda, x = m), \tag{6.3a}$$

$$= \prod_x p(a_x|\lambda, x). \tag{6.3b}$$

We arrive at a useful formula by now going *backwards*, and asking what is the probability that the result of the measurement $\mathbb{M}_x$ will be a. This will be the *total* probability, over all vectors $\vec{a}$, such that the xth component is equal to a, $a_x = a$. In particular, we have

$$p(a|\lambda, x) = \sum_{\vec{a}} \delta_{a_x, a} p(\vec{a}|\lambda). \tag{6.4}$$

That is, we can view $p(a|\lambda, x)$ as the *marginal* of the distribution $p(\vec{a}|\lambda)$, i.e. $p(a|\lambda, x) = p(a_x = a|\lambda)$. To see why this is useful, (6.4) can be substituted into (6.1) to obtain

$$M_{a|x} = \sum_\lambda p(a|\lambda, x) N_\lambda, \tag{6.5a}$$

$$= \sum_\lambda \sum_{\vec{a}} \delta_{a_x, a} p(\vec{a}|\lambda) N_\lambda, \tag{6.5b}$$

$$= \sum_{\vec{a}} \delta_{a_x, a} N'_{\vec{a}} \tag{6.5c}$$

where we have defined $N'_{\vec{a}}$ by

$$N'_{\vec{a}} = \sum_\lambda p(\vec{a}|\lambda) N_\lambda. \tag{6.6}$$

Since $N_\lambda \succcurlyeq 0$ and $p(\vec{a}|\lambda) \geqslant 0$, we see immediately that $N'_{\vec{a}} \succcurlyeq 0$. Moreover, summing over $\vec{a}$ it is seen that

$$\sum_{\vec{a}} N'_{\vec{a}} = \sum_{\vec{a}} \sum_\lambda p(\vec{a}|\lambda) N_\lambda, \tag{6.7a}$$

$$= \sum_\lambda N_\lambda, \tag{6.7b}$$

$$= \mathbb{1}, \tag{6.7c}$$

where the second line follows since $p(\vec{a}|\lambda)$ is a normalised conditional probability distribution, and the third line follows because $\mathbb{N}$ is a normalised measurement. This shows that $\mathbb{N}' = \{N'_{\vec{a}}\}_{\vec{a}}$ is a valid measurement, with outcomes $\vec{a}$. From this it follows that we can interpret (6.5c) in an interesting way: it says that there is a class of *canonical* parent measurements, whose outcomes are labelled by the string of outcomes of the m measurements $\mathbb{M}_x$. When using these measurements, we *deterministically* assign the outcome $a = a_x$ to the measurement labelled x. Moreover, this canonical measurement has a finite number of outcomes o^m.

In this form, we see that there is a strong parallel between being jointly measurable and the classical marginal problem 1.1, with the set of measurements $\{\mathbb{M}_x\}$ being the (single-body) marginals of the canonical parent measurement $\mathbb{N}$.

Since we can always arrive at a canonical measurement and deterministic assignment starting from any parent measurement $\mathbb{N}$ and any probabilistic assignment $p(a|\lambda, x)$ through (6.6), we see that it is possible to instead search exclusively over canonical parents without any loss of generality, and that the feasibility problem (6.2) can be reformulated as the following SDP

$$\text{find} \quad \mathbb{N} \tag{6.8a}$$

$$\text{subject to} \quad M_{a|x} = \sum_{\vec{a}} \delta_{a_x, a} N_{\vec{a}} \qquad \forall\, a, x \tag{6.8b}$$

$$N_{\vec{a}} \succcurlyeq 0 \qquad \forall\, \vec{a}, \tag{6.8c}$$

$$\sum_{\vec{a}} N_{\vec{a}} = \mathbb{1}. \tag{6.8d}$$

In this form all constraints are linear. It is important to note here that the number of constraints grows *exponentially* with the number of measurements m, since (6.8c) contains o^m positivity constraints—one for each element of the exponentially big parent measurement. This means solving this SDP numerically becomes quickly impractical when m grows.

6.2 Two dichotomic measurements

As a more concrete example, we will now study in detail the simplest situation that is interesting from the perspective of measurement incompatibility—when there are two measurements, each of which has two outcomes, which are often referred to as *dichotomic* measurements.

We will begin by going beyond the feasibility problem, and consider how to relax (6.8) into an optimisation form. The method that we will adopt is similar to the one used in the context of measurement informativeness, when we considered a robustness-like measure. But here we will follow the same approach as in the case of the random robustness of entanglement, and consider, as the noise, a completely trivial two-outcome measurement which has POVM elements $T_a = \frac{1}{2}\mathbb{1}$ for $a = 1, 2$. Such a measurement will return a uniformly at random, independent of the quantum state being measured. We now imagine the trivial situation, where both measurements are *identical* and equal to this trivial measurement, then such a pair of measurement is—by construction—jointly measurable. If we treat this as a type of noise, it follows that for any set of measurement $\{\mathbb{M}_x\}$, we can consider noisy versions, with POVM elements equal to

$$M'_{a|x} = \frac{M_{a|x} + r\frac{\mathbb{1}}{2}}{1 + r} \qquad \forall a, x. \tag{6.9}$$

For sufficiently large r, this set of measurements is guaranteed to be jointly measurable. The smallest r that we can take, such that the set $\{\mathbb{M}'_x\}$ is jointly measurable, is therefore a *quantifier* of how incompatible the original set of measurements $\{\mathbb{M}_x\}$ is; if the set is already jointly measurable, then no noise needs to be added; on the other hand, if the set needs a lot of noise to be added before it becomes jointly measurable, then the set of measurements is naturally very incompatible. This type of quantifier is known as the *random robustness of measurement incompatibility*. Here the word 'random' distinguishes that the noise is fixed to be the set of trivial measurements mentioned before, in contrast to the *generalised robustness*, where the noise was a general measurement

We can use the random robustness to relax the feasibility problem (6.8). In particular, we arrive at the minimisation problem

$$\text{minimise} \quad r \tag{6.10a}$$

$$\text{subject to} \quad \frac{M_{a|x} + r\frac{\mathbb{1}}{2}}{1 + r} = \sum_{\vec{a}} \delta_{a_x, a} N_{\vec{a}} \quad \forall a, x, \tag{6.10b}$$

$$N_{\vec{a}} \geqslant 0 \qquad \forall \vec{a}, \tag{6.10c}$$

$$\sum_{\vec{a}} N_{\vec{a}} = \mathbb{1}, \tag{6.10d}$$

$$r \geqslant 0. \tag{6.10e}$$

We note that since two dichotomic measurements are being considered here, the parent measurement has only $2^2 = 4$ outcomes, $\mathbb{N} = \{N_{1,1}, N_{1,2}, N_{2,1}, N_{2,2}\}$. One potential complication that we have introduced is that the problem is no longer an SDP as written. This can however easily be rectified by making a small change of variables. In particular, by defining $\widetilde{N}_{\vec{a}} = (1 + r)N_{\vec{a}}$. This represents an *unnormalised* parent measurement, since it satisfies the normalisation-type condition

$$\sum_{\vec{a}} \widetilde{N}_{\vec{a}} = (1 + r)\mathbb{1}. \tag{6.11}$$

In terms of unnormalised parent measurements, the random robustness is indeed seen to be an SDP, given by

$$\begin{aligned} &\text{minimise} && r && (6.12a)\\ &\text{subject to} && M_{a|x} + r\frac{\mathbb{1}}{2} = \sum_{\vec{a}} \delta_{a_x, a} \tilde{N}_{\vec{a}} \quad \forall a, x, && (6.12b)\\ &&& N_{\vec{a}} \geqslant 0 \quad \forall \vec{a}, && (6.12c)\\ &&& \sum_{\vec{a}} N_{\vec{a}} = (1 + r)\mathbb{1}, && (6.12d)\\ &&& r \geqslant 0. && (6.12e) \end{aligned}$$

It will be advantageous to simplify further, and use the equality constraints (6.12b) and normalisation constraint (6.12d) to solve for all but one element of the unnormalised parent measurement. This is carried out in exercise 6.1 below. There it is shown that we arrive at the following equivalent SDP:

$$\begin{aligned} &\text{minimise} && r && (6.13a)\\ &\text{subject to} && M_{1|1} + r\frac{\mathbb{1}}{2} \geqslant \tilde{N}_{1,1}, && (6.13b)\\ &&& M_{1|2} + r\frac{\mathbb{1}}{2} \geqslant \tilde{N}_{1,1}, && (6.13c)\\ &&& \tilde{N}_{1,1} \geqslant M_{1|2} - M_{2|1}, && (6.13d)\\ &&& \tilde{N}_{1,1} \geqslant 0, && (6.13e)\\ &&& r \geqslant 0. && (6.13f) \end{aligned}$$

We will continue studying this SDP in the 'Advanced topics' section 6.5.1. It will be seen there that by studying its dual formulation, we uncover an interesting and surprising quantitative connection to quantum nonlocality.

We will end this chapter by looking at the simplest example of measurement incompatibility—a pair of Pauli measurements.

Example 6.1. Random robustness of Pauli X and Z measurements.
In this example we will explicitly solve the above SDP in order to find the random robustness of Pauli X and Z measurements. In particular, this means we consider

$$M_{1|1} = |+\rangle\langle+| = \frac{\mathbb{1} + X}{2}, \qquad M_{1|2} = |0\rangle\langle 0| = \frac{\mathbb{1} + Z}{2}, \tag{6.14}$$

$$M_{2|1} = |-\rangle\langle-| = \frac{\mathbb{1} - X}{2}, \qquad M_{2|2} = |1\rangle\langle 1| = \frac{\mathbb{1} - Z}{2}, \tag{6.15}$$

with $X = |+\rangle\langle+| - |-\rangle\langle-| = |0\rangle\langle 1| + |1\rangle\langle 0|$ and $Z = |0\rangle\langle 0| - |1\rangle\langle 1|$.

It will be convenient to write $\widetilde{N}_{1,1}$ in Bloch-vector notation, i.e. to write

$$\widetilde{N}_{1,1} = \frac{\mu \mathbb{I} + \vec{\nu} \cdot \vec{\sigma}}{2}, \tag{6.16}$$

where $\vec{\sigma} = (X, Y, Z)$ is the vector of Pauli operators. This is convenient, since the eigenvalues of any operator of the form $\vec{n} \cdot \vec{\sigma}$ are $\lambda_{\pm} = \pm\|\vec{n}\|_2$.

We will restrict our attention here to parent measurements $\mathbb{N}$ that have rank-1 POVM elements. Such measurements are *extremal*, and lie on the boundary of the set of measurements. An optimal parent measurement will be extremal within the feasible set of the SDP (as it true for all SDPs), however, due to the constraints of the problem, it is not guaranteed that the feasible set contains rank-1 measurements. It is nevertheless a reasonable assumption to try and seek such a solution. Moreover, we will find such a solution if and only if the constraints (6.13b)—(6.13e) are all saturated simultaneously. If we find a parent with rank-1 elements it means that all of the inequalities can indeed be saturated simultaneously. If on the other hand this weren't possible, a contradiction would necessarily be run into, and therefore we would have to revise the assumption.

With all of this in place, the final constraint (6.13e) in this instance is therefore equivalent to

$$\mu = \|\vec{\nu}\|_2, \tag{6.17}$$

where we have imposed that this constraint is saturated, which implies that $\lambda_- = 0$ and $\widetilde{N}_{1,1}$ is a rank-1 operator.

Moving now to the first constraint (6.13b), after substituting both (6.16) and $\mu = \|\vec{\nu}\|_2$, and simplifying, we arrive at

$$(1 + r - \mu)\mathbb{I} + (1 - \nu_1, \nu_2, \nu_3) \cdot \vec{\sigma} \succcurlyeq 0. \tag{6.18}$$

Using the fact that the eigenvalues of $(1 - \nu_1, \nu_2, \nu_3) \cdot \vec{\sigma}$ are $\lambda_{\pm} = \pm\sqrt{1 - 2\nu_1 + \|\vec{\nu}\|_2^2}$, after further rearrangement and simplification we arrive at

$$\nu_1 = \frac{1}{2} + (1 + r)\left(\mu - \frac{1}{2}(1 + r)\right). \tag{6.19}$$

By symmetry, it also follows immediately that (6.13c) is equivalent to

$$\nu_3 = \frac{1}{2} + (1 + r)\left(\mu - \frac{1}{2}(1 + r)\right), \tag{6.20a}$$

$$= \nu_1. \tag{6.20b}$$

Finally, applying the same technique, (6.13d) is readily shown to be equivalent to

$$\nu_1 + \nu_3 = 1. \tag{6.21}$$

Given that in (6.20) we saw that $\nu_1 = \nu_3$, it immediately follows that

$$\nu_1^* = \frac{1}{2}, \qquad \nu_3^* = \frac{1}{2}, \tag{6.22}$$

and therefore from (6.20) that

$$r = 2\mu - 1. \tag{6.23}$$

Our goal is to minimise r. The only freedom that remains is in ν_2, and the minimum value of r is easily seen to occur when $\nu_2^* = 0$, in which case $\mu = \|\nu\|_2$ is minimised, and equal to $\mu^* = \frac{1}{\sqrt{2}}$. Thus, we finally find

$$r^* = \sqrt{2} - 1. \tag{6.24}$$

It is most insightful not to look at this value of robustness, but instead to look at the parent measurement that achieves this robustness. After substituting everything back in, we find that

$$N_{1,1}^* = \frac{1}{2}\frac{\mathbb{I} + \frac{X+Z}{\sqrt{2}}}{2}, \qquad N_{1,2}^* = \frac{1}{2}\frac{\mathbb{I} + \frac{X-Z}{\sqrt{2}}}{2}, \tag{6.25a}$$

$$N_{2,1}^* = \frac{1}{2}\frac{\mathbb{I} - \frac{X-Z}{\sqrt{2}}}{2}, \qquad N_{2,2}^* = \frac{1}{2}\frac{\mathbb{I} - \frac{X+Z}{\sqrt{2}}}{2}. \tag{6.25b}$$

This optimal parent measurement has a natural interpretation: with probability 1/2, a measurement of the observable $\frac{X+Z}{\sqrt{2}}$ is made, while with probability 1/2, $\frac{X-Z}{\sqrt{2}}$ is measured. These measurements can be thought of as trying to measure 'superpositions' of X and Z. If the first measurement is performed, and the outcome +1 is observed, then the results for the X and Z measurements are set to $(1, 1)$, since the +1 eigenstate of $\frac{X+Z}{\sqrt{2}}$ has a large overlap with the +1 eigenstates of both X and Z. Similarly, if the result −1 is observed, then the X and Z measurement results are set to $(-1, -1)$, for the same reason. In this way, although we do not manage to reproduce the statistics of X and Z exactly, their 'noisy' versions are nevertheless reproduced. In particular, the first child measurement of this parent has POVM elements

$$M'_{1|1} = N_{1,1} + N_{1,2} = \frac{\mathbb{I} + \frac{X}{\sqrt{2}}}{2}, \qquad M'_{2|1} = N_{2,1} + N_{2,2} = \frac{\mathbb{I} - \frac{X}{\sqrt{2}}}{2}. \tag{6.26}$$

This can be considered a noisy-X measurement since it can be thought of as performing a measurement of X with probability $\frac{1}{\sqrt{2}}$ and producing a uniformly random measurement outcome with probability $1 - \frac{1}{\sqrt{2}}$.

We end by noting that this example highlights why joint measurability is a more meaningful notion than commutativity for general (non-projective) measurements. As can be easily seen, the noisy child measurements that the above parent leads to do not commute. From the perspective of commutation, it would appear therefore that they are not compatible. However, this conclusion is wrong, since they can be jointly measured, by performing the above parent measurement.

Exercises

6.1 In this exercise, we will derive the SDP for the random robustness of two dichotomic measurements as given in (6.13).

(a) Considering first the case $\vec{a} = (1, 1)$, show that the constraint (6.12b) implies that

$$\widetilde{N}_{1,2} = M_{1|1} + r\frac{\mathbb{I}}{2} - \widetilde{N}_{1,1}. \tag{6.27}$$

(b) Similarly, considering the case $\vec{a} = (1, 2)$, show that the constraint (6.12b) implies that

$$\widetilde{N}_{2,1} = M_{1|2} + r\frac{\mathbb{1}}{2} - \widetilde{N}_{1,1}. \tag{6.28}$$

(c) Verify that the cases $\vec{a} = (2, 1)$ and $\vec{a} = (2, 2)$ do not lead to any new equations. Explain why this is the case.

Hint: You will need to use the normalisation conditions for the measurements $\mathbb{M}_x$ *and* $\widetilde{\mathbb{N}}$.

(d) Use the normalisation condition (6.12d), in conjunction with (6.27) and (6.28) to show that

$$\widetilde{N}_{2,2} = \widetilde{N}_{1,1} - M_{1|2} + M_{2|1} \tag{6.29}$$

(e) Using your answers to parts (a), (b) and (d), and the fact that $N_{\vec{a}} \succcurlyeq 0$, show that the SDP (6.12) can be re-expressed as (6.13).

6.3 Concluding remarks

In this brief chapter we have studied measurement incompatibility in the most general setting, considering arbitrary POVM measurements. The one key take-home message is the following:

- **Measurement incompatibility.** The problem of determining if a set of measurements can be performed simultaneously can be determined by an SDP—see equation (6.8).

6.4 Further reading

- Heinosaari T, Miyadera T and Ziman M 2016 An invitation to quantum incompatibility *J. Phys. A: Math. Theor.* **49** 123001

6.5 Advanced topics

6.5.1 Measurement incompatibility and quantum nonlocality

In this section, we will present a surprising result, obtained through SDP duality: a pair of dichotomic measurements is incompatibile *if and only if they can be used to violate the CHSH Bell inequality*. That is, if they are incompatible, then there is a joint entangled state that can be shared between two parties, and a pair of dichotomic measurements for the second party, such that the statistics generated violate the CHSH Bell inequality. Conversely, if the measurements are jointly measurable, then it is well known that they can never lead to nonlocality, since both *entanglement* and *incompatible measurements* are necessary in order to produce nonlocality.

Our starting point is to look at the dual formulation of (6.13), the SDP we found previously for determining whether a pair of dichotomic measurements is compatible or not. In exercise 6.2 it is shown that the dual of this SDP is given by

$$\text{maximise} \quad \mathrm{tr}(YM_{1|2}) - \mathrm{tr}(WM_{1|1}) - \mathrm{tr}(XM_{1|2}) - \mathrm{tr}(YM_{2|1}) \tag{6.30a}$$

$$\text{subject to} \quad W + X = Y + Z, \tag{6.30b}$$

$$\mathrm{tr}(W + X) = 2, \tag{6.30c}$$

$$W \succcurlyeq 0, \quad X \succcurlyeq 0, \quad Y \succcurlyeq 0, \quad Z \succcurlyeq 0. \tag{6.30d}$$

In order to appreciate the significance of the dual formulation, we need to understand the significance of the dual variables. It was shown previously in exercise 4.12 that an ensemble can be specified by a collection of subnormalised states, that sum up to a normalised state. If we therefore consider $\sigma_{1|1} = W/2$ and $\sigma_{2|1} = X/2$, these can be interpreted as specifying an ensemble $\widetilde{\mathcal{E}}_1 = \{\sigma_{1|1}, \sigma_{2|1}\}$. Similarly, we could consider $\sigma_{1|2} = Y/2$ and $\sigma_{2|2} = Z/2$, as specifying a second ensemble $\widetilde{\mathcal{E}}_2 = \{\sigma_{1|2}, \sigma_{2|2}\}$. The constraint (6.30b) then says something interesting—it says that these two ensembles lead to the *same average density operator*.

Crucially, whenever we have two (or more) ensembles that have the same average density operator, then they can always be created by performing appropriate measurements on the *same pure entangled state*. That is, there are a pair of measurements $\mathbb{Q}_1 = \{Q_{1|1}, Q_{2|1}\}$ and $\mathbb{Q}_2 = \{Q_{1|2}, Q_{2|2}\}$ and a bipartite quantum state $|\psi\rangle$ such that

$$\sigma_{a|x} = \mathrm{tr}_B\Big[(\mathbb{1} \otimes Q_{a|x})|\psi\rangle\langle\psi|\Big]. \tag{6.31}$$

What this shows is that the optimisation in (6.30) can be interpreted as optimising over the measurements $\mathbb{Q}_1$ and $\mathbb{Q}_2$ and the state $|\psi\rangle$. The final step is to re-express the objective function. We see, for example, that

$$\mathrm{tr}(YM_{1|2}) = 2\mathrm{tr}(\sigma_{1|2}M_{1|2}) = 2\mathrm{tr}(\mathrm{tr}_B\Big[(\mathbb{1} \otimes Q_{1|2})|\psi\rangle\langle\psi|\Big]M_{1|2}), \tag{6.32a}$$

$$= 2\mathrm{tr}\Big[(M_{1|2} \otimes Q_{1|2})|\psi\rangle\langle\psi|\Big], \tag{6.32b}$$

$$= 2\langle\psi|(M_{1|2} \otimes Q_{1|2})|\psi\rangle \tag{6.32c}$$

The final form is a *conditional joint probability*—the probability of the pair of outcomes (1, 1) when the measurements $\mathbb{M}_2$ and $\mathbb{Q}_2$ are performed on the first and second systems, respectively, of the bipartite state $|\psi\rangle$. These probabilities are precisely those which occur in a *Bell test*, where distant measurements are performed on a shared quantum state. We therefore arrive at the final form for (6.30), namely

$$\text{maximise} \quad 2\langle\psi|(M_{1|2} \otimes Q_{1|2} - M_{1|1} \otimes Q_{1|1} - M_{1|2} \otimes Q_{2|1} - M_{2|1} \otimes Q_{1|2})|\psi\rangle \tag{6.33a}$$

$$\text{subject to} \quad Q_{1|1} + Q_{2|1} = \mathbb{1}, \tag{6.33b}$$

$$Q_{1|1} + Q_{2|1} = \mathbb{1}, \tag{6.33c}$$

$$Q_{1|1} \succcurlyeq 0, \quad Q_{2|1} \succcurlyeq 0, \quad Q_{1|2} \succcurlyeq 0, \quad Q_{2|2} \succcurlyeq 0, \tag{6.33d}$$

$$\langle\psi|\psi\rangle = 1. \tag{6.33e}$$

Although this is no longer written as an SDP, it has an important interpretation. This problem seeks to find the largest violation of the expression

$$\beta = 2[p(1, 1|2, 2) - p(1, 1|1, 1) - p(1, 2|2, 1) - p(2, 1|1, 2)] \tag{6.34}$$

where $p(a, b|x, y) = \langle\psi|(M_{a|x} \otimes Q_{b|y})|\psi\rangle$ optimised over all shared quantum states $|\psi\rangle$ and over all choices of measurements for Bob $\mathbb{Q}_1$ and $\mathbb{Q}_2$, but with *fixed* measurements for Alice $\mathbb{M}_1$ and $\mathbb{M}_2$. Since strong duality holds, we have $\beta^* = r^*$, and hence $\beta^* > 0$ implies that the measurements $\mathbb{M}_1$ and $\mathbb{M}_2$ are incompatible, and that a strictly positive amount of noisy needs to be added in order to make them jointly measurable.

Although it may not be immediately obvious, the expression (6.34) is a Bell expression—that arises in the study of quantum nonlocality. It is a linear combination of probabilities, all of which can be obtained by performing measurements on a joint entangled state.

In fact (6.34) is something very important indeed. It is the Clauser–Horne–Shimony–Holt (CHSH) Bell inequality in disguise. This form is the so-called 'CH form', with local-bound equal to 0. That is, any local-hidden-variable model that can produce correlations of the form $p(a, b|x, y)$ can never achieve of value greater than 0 in (6.34).

What this shows is that a pair of dichotomic measurements are incompatible if and only if they can violate the CHSH Bell inequality, and can thus be used to demonstrate nonlocality.

Exercises

6.2 In this exercise we will derive the dual SDP of (6.13).

(a) Write down the Lagrangian associated to (6.13), associating dual variables W, X, Y and Z to the four constraints, in order.

(b) Show that the Lagrangian lower bounds the primal objective value if all the dual variables are positive semidefinite.

(c) Show that the Lagrangian is independent of the primal variables if

$$W + X = Y + Z, \qquad \text{tr}(W + Z) = 2. \tag{6.35}$$

(d) Show therefore that the dual SDP is given by

$$\text{maximise} \quad \text{tr}(YM_{1|2}) - \text{tr}(WM_{1|1}) - \text{tr}(XM_{1|2}) - \text{tr}(YM_{2|1}) \tag{6.36a}$$

$$\text{subject to} \quad W + X = Y + Z, \tag{6.36b}$$

$$\text{tr}(W + Z) = 2, \tag{6.36c}$$

$$W \succcurlyeq 0, \quad X \succcurlyeq 0, \quad Y \succcurlyeq 0, \quad Z \succcurlyeq 0. \tag{6.36d}$$

(e) Show that both the primal and dual SDPs are strictly feasible, and hence that strong duality holds.

IOP Publishing

Paul Skrzypczyk and Daniel Cavalcanti

Chapter 7

Quantum channels

In the last chapter of this book we will study *quantum channels*—describing the general evolution of a quantum state. We will see here, as in previous chapters, that the tool of semidefinite programming (SDP) can be used to study a wide range of problems involving quantum channels. The main ingredient that facilitates this is the so-called *Choi–Jamiołkowski* isomorphism, which provides a mapping between quantum channels and bipartite quantum states, which are then positive semidefinite operators, and facilitate the use of SDP techniques.

We will begin by first summarising the key features of the Choi–Jamiołkowski isomorphism before considering a number of simple applications. As in the previous chapters on quantum states and quantum measurements, the most basic application is that of *channel estimation*—which is often referred to as *process tomography*.

We will also show how duality can lead to some surprising results. In particular, we will show that a natural question concerning quantum channels—to find the maximal overlap of a quantum state when applying local quantum channels on one side—in fact is given by the *conditional min-entropy*, a well-known quantity in quantum information with applications in quantum cryptography. We will also see how an important distinguishability measure on quantum channels—the *diamond norm*—can be cast as an SDP, which then allows for many problems involving approximation to be solved by SDP.

Finally, we will also see that a slightly more general object—a collection of *sub-channels* (also often referred to as an *instrument*)—which captures probabilistic transformations, i.e. post-measurement states, can also be studied using SDPs, and outline a couple of simple examples to demonstrate this.

7.1 The Choi–Jamiołkowski isomorphism

The central ingredient that we need in order to use SDP to study problems involving quantum channels is the Choi–Jamiołkowski isomorphism. Recall that a quantum

doi:10.1088/978-0-7503-3343-6ch7 © IOP Publishing Ltd 2023

channel is a linear map from density operators to density operators, which is both *trace preserving* and *completely positive*. That is, a channel $\Lambda(\cdot)$ will map the density operator ρ to another density operator $\sigma = \Lambda(\rho)$. Trace preservation ensures that σ has $\mathrm{tr}(\sigma) = 1$, as required to be a valid density operator. σ also needs to be a positive operator. However, imposing that $\Lambda(\cdot)$ transforms positive matrices into positive matrices is not enough: if ρ is actually the reduced state of a larger state ρ_{AB} such that $\rho = \mathrm{tr}_B(\rho_{AB})$, then we should also impose that the final (bipartite) state is positive semidefinite, i.e.

$$\Lambda \otimes \mathrm{id}(\rho_{AB}) = \sigma_{AB} \geqslant 0 \quad \text{for all } \rho_{AB}, \tag{7.1}$$

where $\mathrm{id}(\cdot)$ is the *identity channel*, that leaves all states unchanged, $\mathrm{id}(\omega) = \omega$, for all ω, and the dimension of B is arbitrary[1]. If a channel satisfies (7.1), we say that it is *completely positive*.

The Choi–Jamiołkowski isomorphism states that there is a one-to-one mapping between every quantum channel, and a subset of bipartite quantum states. In particular, to any given channel $\Lambda(\cdot)$, we can associate the so-called *Choi state* X_Λ, through the relationship

$$X_\Lambda = \Lambda \otimes \mathrm{id}(|\Phi^+\rangle\langle\Phi^+|), \tag{7.2}$$

where $|\Phi^+\rangle = \frac{1}{\sqrt{d}}\sum_i |i\rangle|i\rangle$ is the maximally entangled state (of two qudits). Because $\Lambda(\cdot)$ is trace preserving and completely positive, X_Λ is a valid density operator, for any channel $\Lambda(\cdot)$, and therefore satisfies $X_\Lambda \geqslant 0$ and $\mathrm{tr}(X_\Lambda) = 1$. We can see however that not all states can arise as Choi states. In particular, if we consider tracing out the first subsystem, because the channel only acts locally on this system, the reduced density operator of the second system must remain unchanged, and therefore be the maximally mixed state, that is

$$\mathrm{tr}_A(X_\Lambda) = \frac{1}{d}\mathbb{1}. \tag{7.3}$$

The Choi–Jamiołkowski isomorphism says that the set of Choi states is in one-to-one correspondence with the set of quantum channels. That is, given a density operator X, satisfying $\mathrm{tr}_A(X) = \frac{1}{d}\mathbb{1}$, then this uniquely specifies a completely positive and trace-preserving quantum channel. The associated channel $\Lambda_X(\cdot)$ is

$$\Lambda_X(\rho) = d\,\mathrm{tr}_B[(\mathbb{1} \otimes \rho^\top)X], \tag{7.4}$$

where the transpose (which is basis-dependent) is taken *in the basis of the maximally entangled state* used to define X.

[1] Note that, although arbitrary, it suffices to check only for $d_B = d_A$, due to the fact that it is ultimately entanglement between the two systems that is important, and the maximally-sized subspace of B that can be entangled with A is precisely of dimension d_B.

Exercises

7.3 In this exercise, we will confirm that if we use the Choi state X_Λ of a channel $\Lambda(\cdot)$, then the channel $\Lambda_{X_\Lambda}(\cdot)$ defined through (7.4) is indeed just $\Lambda(\cdot)$, as required.

(a) By substituting the definition of the maximally entangled state, show that the Choi state can alternatively be written as

$$X_\Lambda = \frac{1}{d}\sum_{i,j}\Lambda(|i\rangle\langle j|) \otimes |i\rangle\langle j|. \tag{7.5}$$

(b) Substitute X_Λ (in the form from (7.5)) into (7.4), to show that $\Lambda_{X_\Lambda}(\cdot) = \Lambda(\cdot)$.

7.4 In this exercise we will study another representation of quantum channels, in terms of so-called *Kraus operators.*

(a) Show that any map of the form

$$\Lambda(\rho) = \sum_a K_a \rho K_a^\dagger, \tag{7.6}$$

is completely positive, for any set of operators $\{K_a\}$, for $a = 1, \ldots, N$. *The operators K_a are called Kraus operators, and this decomposition is known as the Kraus decomposition of the map.*

(b) Show that in order to be trace preserving the Kraus operators must satisfy

$$\sum_a K_a^\dagger K_a = \mathbb{1}. \tag{7.7}$$

(c) Combining (7.2) and the Kraus decomposition (7.6), show that the Choi state can be written as

$$X_\Lambda = \frac{1}{d}\sum_{i,j,a} K_a|i\rangle\langle j|K_a^\dagger \otimes |i\rangle\langle j|. \tag{7.8}$$

7.5 In this exercise we will consider a special class of completely positive maps called *unital maps.* Consider the dual map $\Lambda^\dagger(\cdot)$ of a completely positive trace preserving map $\Lambda(\cdot)$, that is, the unique map that satisfies

$$\text{tr}[M\Lambda(\rho)] = \text{tr}[\Lambda^\dagger(M)\rho], \tag{7.9}$$

for all operators ρ and M.

(a) Show that $\Lambda^\dagger(\cdot)$ is completely positive if and only if $\Lambda(\cdot)$ is completely positive.

(b) Show that $\Lambda^\dagger(\cdot)$ satisfies the *unitality condition*

$$\Lambda^\dagger(\mathbb{1}) = \mathbb{1}, \tag{7.10}$$

if and only if $\Lambda(\cdot)$ is trace preserving.
The physical significance of this is that the dual map is also normalisation-preserving, but now in the Heisenberg representation. In particular, this condition ensures that a valid (normalised) measurement remains normalised after application of the (dual) channel $\Lambda^\dagger(\cdot)$. That is, $\mathbb{M}' = \{M_a'\}$ with $M_a' = \Lambda^\dagger(M_a)$ is a valid POVM whenever $\mathbb{M} = \{M_a\}$ is a valid POVM.

(c) Show that if a map $\Lambda^\dagger(\cdot)$ is completely positive and unital, then the corresponding Choi state satisfies

$$X_{\Lambda^\dagger} \succcurlyeq 0, \qquad \mathrm{tr}_B(X_{\Lambda^\dagger}) = \frac{1}{d}\mathbb{1}. \tag{7.11}$$

7.6 In this exercise we will consider more general maps than quantum channels, i.e. linear maps which need not be either completely positive nor trace preserving. We will however require all maps to preserve hermiticity, i.e. to keep the output Hermitian (since the input will always be assumed to be a density operator, which is therefore Hermitian).

(a) Show that if a map $\Gamma(\cdot)$ is trace non-increasing, meaning that $\mathrm{tr}[\Gamma(\sigma)] \leqslant \mathrm{tr}[\sigma]$ for all states σ, then the associated Choi state will satisfy

$$\mathrm{tr}_A(X_\Gamma) \preccurlyeq \frac{1}{d}\mathbb{1}. \tag{7.12}$$

(b) Show that if a map $\Gamma(\cdot)$ is can be written as the difference of two quantum channels, $\Gamma(\cdot) = \Lambda(\cdot) - \Omega(\cdot)$, then

$$\mathrm{tr}_A(X_\Gamma) = 0. \tag{7.13}$$

We now have enough background in place in order to show how the Choi–Jamiołkowski isomorphism allows for problems involving quantum channels to be cast as SDPs. We will make this explicit through the use of an example, that of channel estimation.

7.2 Channel estimation

In this section we will consider a natural problem, that of channel estimation, which is the analogue of state estimation from chapter 3 and measurement estimation from chapter 4. Consider a situation where one has access to a channel, which is yet to be characterised. The only knowledge about the channel that one has is its action on a fixed set of input states. That is, one has access to the data $\{\rho_i, \sigma_i\}$ for $i = 1, \ldots, N$, such that $\Lambda(\rho_i) = \sigma_i$. The goal is to determine whether or not this data is consistent with a completely positive and trace preserving channel (or possibly a set of channels), and to find a channel able to reproduce the results, if it exists. This is therefore an instance of a feasibility problem, which can be written as

$$\text{find} \quad \Lambda(\cdot) \tag{7.14a}$$

$$\text{subject to} \quad \Lambda(\rho_i) = \sigma_i \qquad i = 1, \ldots, N, \tag{7.14b}$$

$$\mathrm{tr}[\Lambda(\omega)] = 1 \qquad \text{for all } \omega, \tag{7.14c}$$

$$\Lambda \otimes \mathrm{id}(\omega_{AB}) \succcurlyeq 0 \qquad \text{for all } \omega_{AB}. \tag{7.14d}$$

As written, this feasibility problem may appear to be rather complicated, due to the final two constraints—enforcing trace preservation and complete positivity—which must hold for all states, either on the space on which the channel acts, or on all bipartite states. However, using the Choi–Jamiołkowski isomorphism from the previous section, we see that the problem can be recast in terms of the set of Choi states (of the unknown channel $\Lambda(\cdot)$), which turns the two constraints into SDP constraints. In particular, we can recast (7.14) as the following SDP

$$\text{find} \quad X \tag{7.15a}$$

$$\text{subject to} \quad d\,\mathrm{tr}_B[(\mathbb{1} \otimes \rho_i^{\mathsf{T}})X] = \sigma_i \qquad i = 1, \ldots, N, \tag{7.15b}$$

$$\mathrm{tr}_A(X) = \frac{1}{d}\mathbb{1}, \tag{7.15c}$$

$$X \succcurlyeq 0. \tag{7.15d}$$

In this formulation, we have rephrased the three constraints (7.14b)–(7.14d) in terms of Choi states in (7.15b)–(7.15d), all of which are now linear equality or inequality constraints.

As is often the case, it is interesting and important to relax this problem to an optimisation form, rather than the natural feasibility form. A natural way to do this is to allow the constraints to be relaxed. Here, we could seek to find the channel which is best able to produce the target output states σ_i given the input states ρ_i. That is, one possible relaxation is

$$\text{minimise} \quad \delta \tag{7.16a}$$

$$\text{subject to} \quad \left\|\Lambda(\rho_i) - \sigma_i\right\|_1 \leqslant \delta \qquad i = 1, \ldots, N, \tag{7.16b}$$

$$\mathrm{tr}[\Lambda(\omega)] = 1 \qquad \text{for all } \omega, \tag{7.16c}$$

$$\Lambda \otimes \mathrm{id}(\omega_{AB}) \succcurlyeq 0 \qquad \text{for all } \omega_{AB}, \tag{7.16d}$$

where we seek to minimise the largest discrepancy between the output of the channel on any of the input states ρ_i and the target σ_i, as quantified by the natural distance measure on states—the trace distance. As shown in exercise 7.7, this can be cast as the following SDP

$$\text{minimise} \quad \delta \tag{7.17a}$$

$$\text{subject to} \quad \mathrm{tr}(Y_i) \leqslant \delta \qquad i = 1, \ldots, N, \tag{7.17b}$$

$$-Y_i \preccurlyeq d\,\mathrm{tr}_B[(\mathbb{1} \otimes \rho_i^{\mathsf{T}})X] - \sigma_i \preccurlyeq Y_i \qquad i = 1, \ldots, N, \tag{7.17c}$$

$$Y_i \succcurlyeq 0, \qquad i = 1, \ldots, N, \tag{7.17d}$$

$$\mathrm{tr}_A(X) = \frac{1}{d}\mathbb{1}, \qquad X \succcurlyeq 0. \tag{7.17e}$$

Exercises

7.7 Use the SDP formulation of the trace norm from (2.8) to show that (7.16) can be cast as the SDP (7.17).

7.8. In this exercise we will consider a relaxed version of (7.14). Instead of knowing exactly the output of the channel when the input state is ρ_i, it is natural to assume that it is known up to some precision. This can be modelled by enforcing that the output is not equal to σ_i, but only close to it in trace distance, i.e. we have the constraints

$$\|\Lambda(\rho_i) - \sigma_i\|_1 \leqslant \epsilon, \tag{7.18}$$

where ϵ quantifies the uncertainty of the output of the channel, and should be considered as input data into the problem.

Using the SDP characterisation of the ϵ trace norm ball from exercise 2.17, show that the relaxed feasibility problem can be written as

$$\text{find} \quad X \tag{7.19a}$$

$$\text{subject to} \quad d\,\text{tr}_B[(\mathbb{1} \otimes \rho_i^{\mathsf{T}})X] - \sigma_i = Z_i - Y_i \qquad i = 1, \ldots, m, \tag{7.19b}$$

$$\text{tr}(Z_i - Y_i) = \epsilon \qquad i = 1, \ldots, N, \tag{7.19c}$$

$$Z_i \succcurlyeq 0, \quad Y_i \succcurlyeq 0, \qquad i = 1, \ldots, N, \tag{7.19d}$$

$$\text{tr}_A(X) = \frac{1}{d}\mathbb{1}, \tag{7.19e}$$

$$X \succcurlyeq 0. \tag{7.19f}$$

7.9 Explain how exercise 7.8 can be used to find an alternative formulation of the relaxation (7.16), and conversely how (7.17) can equally be use to find an alternative SDP formulation applicable for exercise 7.8.

7.3 The diamond norm and channel discrimination

In this section we will now consider the most natural norm when considering quantum channels, known as the *diamond norm*, and the associated distance measure between quantum channels. This norm can be defined operationally, in analogy to how the trace norm inherits an operational interpretation through the trace distance. This is how we will introduce it here.

Recall that, as seen in section 4.2.2, when considering binary state discrimination, it is the trace distance $T(\omega, \sigma) = \frac{1}{2}\|\omega - \sigma\|_1$ that determines the probability of distinguishing between ω and σ, when performing an optimal measurement for state discrimination. We saw there that $P^*_{\text{guess}} = \frac{1}{2}[1 + T(\omega, \sigma)]$. This showed us that the trace distance, and therefore trace norm, were not only mathematically meaningful definitions, but physically and operationally motivated ones. In order to arrive at a good norm for channels, we will use this as the key inspiration.

With this in mind, we can therefore consider the task of channel discrimination, whereby we are given either the channel $\Lambda(\cdot)$ or $\Omega(\cdot)$ (with equal probability), and wish to guess, as best as possible, which channel was given. Whereas in state discrimination all that could be done was to measure the states, in channel discrimination we need to consider both the state ρ that will be input into the channel, *and* the (binary) measurement $\mathbb{M} = \{(\mathbb{I} + Z)/2, (\mathbb{I} - Z)/2\}$ that will be used to discriminate between the output states. Crucially, we will allow ourselves to send only *part of a bipartite state through the channel*, in order to help distinguish between the two channels as best as possible.

It is helpful to realise, given the way the problem has been formulated above, that we can think of channel discrimination in two parts; first, for a fixed input state ρ, we end up with a binary state discrimination problem between the two states $\Lambda \otimes \mathrm{id}(\rho)$ and $\Omega \otimes \mathrm{id}(\rho)$; second, since we can choose the input state, this varies the binary state discrimination problem (which concerns the output states of the channel), and so to optimally discriminate the channels, we can optimise over all binary state discrimination tasks that the two channels give us access to.

Given the above, we can define the average success probability in correctly guessing which of the two channels was given, in direct analogy to (4.12), by

$$P^*_{\text{guess}} = \text{maximise} \quad \frac{1}{2} + \frac{1}{4}\mathrm{tr}[Z(\Lambda \otimes \mathrm{id}(\rho) - \Omega \otimes \mathrm{id}(\rho))] \tag{7.20a}$$

$$\text{subject to} \quad -\mathbb{I} \preccurlyeq Z \preccurlyeq \mathbb{I}, \tag{7.20b}$$

$$\rho \succcurlyeq 0, \quad \mathrm{tr}[\rho] = 1, \tag{7.20c}$$

and therefore define the *diamond norm* of a map by

$$\|\Gamma(\cdot)\|_\diamond = \text{maximise} \quad \mathrm{tr}[Z(\Gamma \otimes \mathrm{id}(\rho))] \tag{7.21a}$$

$$\text{subject to} \quad -\mathbb{I} \preccurlyeq Z \preccurlyeq \mathbb{I}, \tag{7.21b}$$

$$\rho \succcurlyeq 0, \quad \mathrm{tr}(\rho) = 1, \tag{7.21c}$$

and associated *diamond norm distance* by

$$D_\diamond(\Lambda(\cdot), \Omega(\cdot)) = \frac{1}{2}\|\Lambda(\cdot) - \Omega(\cdot)\|_\diamond. \tag{7.22}$$

With these definitions, the diamond norm distance, by construction, has the same operational interpretation as the trace distance; it quantifies how well two channels can be distinguished in binary channel discrimination.

As written in (7.21), the diamond norm is not an SDP, since the objective function is a nonlinear function, containing a product of the variables Z and ρ. However, as will be shown in section 7.7, we can in fact recast the diamond norm so that it is given by the following SDP:

$$\|\Gamma(\cdot)\|_\diamond = \text{maximise} \quad \text{tr}[YX_\Gamma] \tag{7.23a}$$

$$\text{subject to} \quad -d(\mathbb{1} \otimes \sigma) \preccurlyeq Y \preccurlyeq d(\mathbb{1} \otimes \sigma), \tag{7.23b}$$

$$\sigma \succcurlyeq 0, \quad \text{tr}(\sigma) = 1, \tag{7.23c}$$

where X_Γ is the Choi state[2] associated to $\Gamma(\cdot)$, and d is the dimension of the Hilbert space on which the channel acts. In exercise 7.10 it is shown that the dual SDP of (7.23) is given by

$$\|\Gamma(\cdot)\|_\diamond = \text{minimise} \quad \mu \tag{7.24a}$$

$$\text{subject to} \quad X_\Gamma = Z_2 - Z_1, \tag{7.24b}$$

$$Z_1 \succcurlyeq 0, \quad Z_2 \succcurlyeq 0, \tag{7.24c}$$

$$\mu\mathbb{1} \succcurlyeq d(\text{tr}_A[Z_1 + Z_2]), \tag{7.24d}$$

where we note that Z_1 and Z_2 are operators on the bipartite Hilbert space $\mathcal{H}_A \otimes \mathcal{H}_B$.[3]

In (7.24) it can be seen that μ is minimised when equal to $\mu^* = d\|\text{tr}_A[Z_1 + Z_2]\|_\infty$, and that this value can always be attained. Therefore, we can equivalently express the diamond norm as

$$\|\Gamma(\cdot)\|_\diamond = \text{minimise} \quad d\|\text{tr}_A[Z_1 + Z_2]\|_\infty \tag{7.25a}$$

$$\text{subject to} \quad X_\Gamma = Z_2 - Z_1, \tag{7.25b}$$

$$Z_1 \succcurlyeq 0, \quad Z_2 \succcurlyeq 0. \tag{7.25c}$$

It is important at this point to make a couple of notes. First, although our initial motivation was to measure the distance between quantum channels, the diamond norm is in fact defined on more general quantum *maps* (as considered in exercise 7.8), i.e. linear operations which act on the space of (Hermitian) operators and produce (Hermitian) operators. In particular, we are interested in maps of the form $\Gamma(\cdot) = \Lambda(\cdot) - \Omega(\cdot)$, which are the difference of two quantum channels. Such maps will in general be neither completely positive nor trace preserving. Nevertheless, the diamond norm is well defined on such maps, and equals the objective value of the (primal or dual) SDPs given.

Conversely, if we do consider a completely positive and trace preserving map $\Lambda(\cdot)$ with corresponding Choi state X_Λ, then it necessarily has unit diamond norm, $\|\Lambda(\cdot)\|_\diamond = 1$. To see this, note first that in the primal SDP (7.23) if $Y = \mathbb{1}_{AB}$ and $\sigma = \mathbb{1}_B/d$ are picked as primal variables, then they can easily be seen to be feasible, satisfying $-\mathbb{1}_{AB} \preccurlyeq Y \preccurlyeq \mathbb{1}_{AB}$, and therefore $\|\Lambda(\cdot)\|_\infty \succcurlyeq \text{tr}(X_\Lambda) = 1$. The inequality

[2] Note that we use 'state', but in general X_Γ might not be normalised, if Γ is not a channel (but merely a map).

[3] In order to avoid over-complicating and overloading the notation, we use A to denote *both* the input and the output spaces a channel, even if the channel changes the dimension of the Hilbert space. If ever any potential confusion could arise by using the same label for both spaces, we will instead use different labels, such as A and A'.

follows since we have no guarantee that this choice of primal variables is optimal. On the other hand, in the dual SDP (7.25) a feasible set of variables is to take $Z_2 = X_\Lambda \succcurlyeq 0$, and $Z_1 = 0$. In this case, $\|\mathrm{tr}_A(Z_2)\|_\infty = \|\mathbb{1}/d\|_\infty = \frac{1}{d}$, and therefore $\|\Lambda(\cdot)\|_\diamond \leqslant 1$, where again the inequality follows due to the dual variables being potentially sub-optimal. However, the lower and upper bounds on the diamond norm of the channel match, and hence we conclude that it must be equal to unity (and furthermore that the variables picked are optimal for both the primal and dual SDPs respectively).

As a second note, we can also use the dual formulation (7.24) to define two interesting sets of maps, both of which can then be optimised over inside SDPs. First of all, we can simply define the *unit diamond norm ball*, i.e. all maps which satisfy $\|\Gamma(\cdot)\|_\diamond \leqslant 1$. In particular, this is given by

$$\mathcal{B}_\diamond = \{\Gamma(\cdot) \mid \|\Gamma(\cdot)\|_\diamond \leqslant 1\}, \tag{7.26a}$$

$$= \{\Gamma(\cdot) \mid X_\Gamma = Z_2 - Z_1,\, Z_1 \succcurlyeq 0,\, Z_2 \succcurlyeq 0,\, \mathbb{1} \succcurlyeq d(\mathrm{tr}_A[Z_1 + Z_2])\}. \tag{7.26b}$$

The second interesting set is the set of quantum channels that are ϵ close to a target quantum channel $\Lambda(\cdot)$. That is, the set $\mathcal{B}_\diamond^{\epsilon,\Lambda}$ of channels $\Omega(\cdot)$ such $\|\Lambda(\cdot) - \Omega(\cdot)\|_\diamond \leqslant \epsilon$. This can similarly be specified as

$$\mathcal{B}_\diamond^{\epsilon,\Lambda} = \{\Omega(\cdot) \mid \|\Lambda(\cdot) - \Omega(\cdot)\|_\diamond \leqslant \epsilon\}, \tag{7.27a}$$

$$= \{\Omega(\cdot) \mid X_\Lambda - X_\Omega = Z_2 - Z_1,\, Z_1 \succcurlyeq 0,\, Z_2 \succcurlyeq 0,\, \epsilon\mathbb{1} \succcurlyeq d(\mathrm{tr}_A[Z_1 + Z_2]), \tag{7.27b}$$

$$X_\Omega \succcurlyeq 0,\, \mathrm{tr}_A(X_\Omega) = \mathbb{1}/d\}.$$

As a simple example of a problem where this set arises is to return to the problem of channel estimation from the previous section. Consider now a variant of channel estimation where we would like to know if there is any channel which approximates a fixed target channel $\Lambda(\cdot)$ and that is also able to reproduce the input–output data given. Here, we can take approximation to mean that the channel sought after should be ϵ close to $\Lambda(\cdot)$ in diamond norm. Formally, the feasibility problem to be solved is therefore

$$\text{find} \quad \Omega(\cdot) \tag{7.28a}$$

$$\text{subject to} \quad \Omega(\rho_i) = \sigma_i \qquad i = 1, \ldots, N, \tag{7.28b}$$

$$\|\Lambda(\cdot) - \Omega(\cdot)\|_\diamond \leqslant \epsilon \tag{7.28c}$$

$$\mathrm{tr}[\Omega(\omega)] = 1 \qquad \text{for all } \omega, \tag{7.28d}$$

$$\Omega \otimes \mathrm{id}(\omega_{AB}) \succcurlyeq 0 \qquad \text{for all } \omega_{AB}. \tag{7.28e}$$

Using the characterisation of $\mathcal{B}_\diamond^{\epsilon,\Lambda}$ from (7.27b), this can be expressed as the following feasibility SDP

$$\text{find} \quad X_\Omega \tag{7.29a}$$

$$\text{subject to} \quad d\,\mathrm{tr}_B[(\mathbb{I} \otimes \rho_i^{\mathsf{T}})X_\Omega] = \sigma_i \qquad i = 1, \ldots, N, \tag{7.29b}$$

$$X_\Lambda - X_\Omega = Z_2 - Z_1, \tag{7.29c}$$

$$Z_1 \succcurlyeq 0, \quad Z_2 \succcurlyeq 0, \tag{7.29d}$$

$$\epsilon\mathbb{I} \succcurlyeq d(\mathrm{tr}_A[Z_1 + Z_2]), \tag{7.29e}$$

$$X_\Omega \succcurlyeq 0, \quad \mathrm{tr}_A(X_\Omega) = \frac{1}{d}\mathbb{I}. \tag{7.29f}$$

Exercises

7.10 In this exercise we will derive the dual formulation (7.24) of the diamond norm.
(a) Write down the Lagrangian associated to the primal SDP (7.23), associating dual variables Z_1 and Z_2 to the left-hand and right-hand constraints in (7.23b), and associating Z_3 and μ to the first and second constraints in (7.23c).
(b) Identify the constraints that need to be satisfied by the dual variables so that the Lagrangian both upper bounds the value of the primal objective function, and is independent of the primal variables Y and σ.
(c) Use part (b) to write down the dual formulation of the diamond norm.
(d) Show that Z_3 is a slack variable, and therefore simplify the dual to arrive at (7.24).

7.11 Find an SDP representation of the *diamond norm ϵ ball*, that is, the set of maps $\mathcal{B}_\diamond^\epsilon = \{\Gamma(\cdot) \mid \|\Gamma(\cdot)\|_\diamond \leqslant \epsilon\}$, analogous to (7.26b).

7.12 The SDP (7.29) can be simplified, by noting that Z_1 is a slack variable. Solve for Z_1 to obtain a simplified SDP formulation of (7.29).

7.13 The feasibility problem (7.28) can be relaxed to an optimisation problem, by finding the minimal ϵ for which there is a solution. Write down the associated SDP formulation of this problem.

7.4 The conditional min-entropy and the singlet fidelity

We now turn our attention to a second problem concerning channels, and show that the duality of SDPs allows us to uncover an operational interpretation of an entropic quantity known as the *conditional min-entropy* which at first sight has no obvious physical or operational interpretation.

The question which we will study is *how close is a given bipartite quantum state ρ_{AB} to being maximally entangled?* In chapter 3 several notions of distances between quantum states were studied. Here we will choose to measure closeness using the fidelity, introduced in section 3.1.2. The fidelity between ρ_{AB} and the maximally entangled state $|\Phi^+\rangle = \frac{1}{\sqrt{d}}\sum_i |i\rangle|i\rangle$ could be measured directly, that is,

$F(\rho_{AB}, |\Phi^+\rangle\langle\Phi^+|) = \langle\Phi^+|\rho_{AB}|\Phi^+\rangle$. However there is an obvious drawback to this—namely the fact that the (local) basis in which the maximally entangled state $|\Phi^+\rangle$ was defined is completely arbitrary. This means we might underestimate how entangled the state is, simply because of a mismatch between the basis used to measure the entanglement in. We therefore might want to consider instead

$$F_{\Phi^+}(\rho_{AB}) = \text{maximise} \quad F\big(\rho_{AB}, (U \otimes V)|\Phi^+\rangle\langle\Phi^+|(U^\dagger \otimes V^\dagger)\big) \tag{7.30a}$$

$$\text{subject to} \quad U^\dagger U = \mathbb{1}, \quad V^\dagger V = \mathbb{1}. \tag{7.30b}$$

We will refer to this quantity as the *singlet fidelity*[4]. That is, to optimise over local unitary operators U and V, in order to change the basis of the maximally entangled state $|\Phi^+\rangle$. However, as proven in exercise 7.14, the maximally entangled state $|\Phi^+\rangle$ has an important *invariance*, namely that for any operator A,

$$(A \otimes \mathbb{1})|\Phi^+\rangle = (\mathbb{1} \otimes A^\mathsf{T})|\Phi^+\rangle, \tag{7.31}$$

where, as always, the transpose is in the basis of the maximally entangled state. Using this fact, in exercise 7.14 it is shown that it isn't necessary to optimise over local unitaries on both particles, but to consider only local unitaries on a single particle, which we take to be the first. Physically, this shows that it is possible to change the basis of both particles of the maximally entangled state just by applying local unitaries on one particle. We can therefore set $V = \mathbb{1}$ without any loss of generality in (7.30), and achieve the same fidelity with the maximally entangled state.

It will be interesting to go one step further, and to allow for more general channels to be applied on the first particle, in order to obtain a *one-sided optimised singlet fidelity*. That is, on top of any unitary transformation U—whose purpose is to change the basis of $|\Phi^+\rangle$—we will moreover consider instead of the singlet fidelity of ρ_{AB}, rather the singlet fidelity of $\Lambda \otimes \mathrm{id}(\rho_{AB})$ optimised over all channels $\Lambda(\cdot)$.

This is rather natural, and it allows for the most general (one-sided) manipulation of the state to be performed. Such a procedure cannot generate entanglement, since it is clearly an LOCC operation (in fact, it doesn't even involve any classical communication). Hence, the utility of such a local processing is to allow for a better assessment of the entanglement contained in the state ρ_{AB}. Since the change of basis unitary U can be incorporated into the channel $\Lambda(\cdot)$, and since we are optimising over channels, the quantity of interest is thus

$$F^{(A)}_{\Phi^+}(\rho_{AB}) = \text{maximise} \quad \langle\Phi^+|\Lambda \otimes \mathrm{id}(\rho_{AB})|\Phi^+\rangle \tag{7.32a}$$

$$\text{subject to} \quad \mathrm{tr}[\Lambda(\omega)] = 1 \qquad \text{for all } \omega, \tag{7.32b}$$

$$\Lambda \otimes \mathrm{id}(\omega_{AB}) \succcurlyeq 0 \qquad \text{for all } \omega_{AB}, \tag{7.32c}$$

---

[4] This is because the so-called *singlet state* $|\Psi^-\rangle = \frac{1}{\sqrt{2}}(|0\rangle|1\rangle - |1\rangle|0\rangle)$ is a maximally entangled state, equivalent to $|\Phi^+\rangle$ up to a local unitary transformation $U \otimes V$. We could thus have picked this as our target maximally entangled state without loss of generality; we do not, in order to make a link with the Choi state of the channel.

where we denote by $F^{(A)}_{\Phi^+}(\rho_{AB})$ the *one-sided optimised singlet fidelity*. We could now proceed as before, and use the Choi state of $\Lambda(\cdot)$ to re-express this problem. However, since the objective function already involves $|\Phi^+\rangle$ (which appears in the definition of the Choi state), there is an alternative route, which turns out to be more useful. In particular, we can notice that the objective function can be written as

$$\langle\Phi^+|\Lambda \otimes \mathrm{id}(\rho_{AB})|\Phi^+\rangle = \mathrm{tr}\Big[(\Lambda \otimes \mathrm{id}(\rho_{AB}))|\Phi^+\rangle\langle\Phi^+|\Big], \tag{7.33a}$$

$$= \mathrm{tr}\Big[\rho_{AB}(\Lambda^\dagger \otimes \mathrm{id}(|\Phi^+\rangle\langle\Phi^+|))\Big], \tag{7.33b}$$

$$= \mathrm{tr}\Big[\rho_{AB}X_{\Lambda^\dagger}\Big], \tag{7.33c}$$

where in the second line we used the definition of the adjoint map $\Lambda^\dagger(\cdot)$ to move the channel from ρ_{AB} to $|\Phi^+\rangle$, and in the third line we have then used the fact that this is precisely the definition of the Choi state for the (adjoint) map $\Lambda^\dagger(\cdot)$.

As was shown in exercise 7.5, the adjoint $\Omega^\dagger(\cdot)$ of quantum channel $\Omega(\cdot)$, is a completely positive *unital* map, i.e. one which has the maximally mixed state as a fixed point, $\Omega^\dagger(\mathbb{1}/d) = \mathbb{1}/d$. In terms of the Choi state, it was also shown that this is equivalent to $\mathrm{tr}_B(X_{\Omega^\dagger}) = \mathbb{1}/d$ (and $X_{\Omega^\dagger} \succcurlyeq 0$, which maintains the complete positivity of $\Omega(\cdot)$). Putting all of these ingredients together, we therefore see that (7.32) can be expressed explicitly as an SDP, in terms of the Choi state of $\Lambda^\dagger(\cdot)$, as follows[5]:

$$\text{maximise} \quad \mathrm{tr}\Big[\rho_{AB}X_{\Lambda^\dagger}\Big] \tag{7.34a}$$

$$\text{subject to} \quad \mathrm{tr}_B[X_{\Lambda^\dagger}] = \frac{1}{d}\mathbb{1}, \tag{7.34b}$$

$$X_{\Lambda^\dagger} \succcurlyeq 0. \tag{7.34c}$$

As we can see, the final form of the SDP, in terms of the Choi state, is remarkably different from the original formulation in terms of the channel. Nevertheless, it is just a re-expression, and one that can be used in practice to compute the optimised singlet fidelity of a state. As we will now see, there is a second advantage of this SDP formulation, namely that its dual formulation provides surprising links to a seemingly unrelated quantities.

As shown in exercise 7.5, the dual formulation of (7.34) is

$$\text{minimise} \quad \frac{1}{d}\mathrm{tr}(Y) \tag{7.35a}$$

$$\text{subject to} \quad \rho_{AB} + Z = Y \otimes \mathbb{1}, \tag{7.35b}$$

$$Z \succcurlyeq 0. \tag{7.35c}$$

[5] Note that, as mentioned before, we will continue to call $X_{\Lambda^\dagger}$ a Choi *state* even when $\Lambda^\dagger(\cdot)$ is not a trace preserving map, meaning that $X_{\Lambda^\dagger}$ is a not normalised state. It is nevertheless positive semidefinite, and in order to avoid overloading the terminology, we stick to the term 'state'.

As is also shown in exercise 7.5, both the primal and dual problems are strictly feasible, so strong duality holds, and this is a dual formulation of the optimised singlet fidelity.

We can now proceed in two ways, both of which are interesting. The first way is to realise that (7.35) can be seen as a type of generalised *robustness*, as has been seen many times before. In particular, we can first note that since $\rho_{AB} \succcurlyeq 0$ (by assumption), and $Z \succcurlyeq 0$, due to (7.35c), it must be the case that $Y \succcurlyeq 0$, since otherwise it is impossible for the equality constraint (7.35b) to hold. We can therefore interpret each of these dual variables as (unnormalised) quantum states and define

$$\omega = \frac{Z}{\mathrm{tr}(Z)}, \qquad \sigma = \frac{Y}{\mathrm{tr}(Y)}. \tag{7.36}$$

If we further define $r = \mathrm{tr}(Z)$, then taking the trace of the equality constraint (7.35b), it is seen that

$$\mathrm{tr}(Y) = \frac{1+r}{d}. \tag{7.37}$$

Altogether this allows us to re-express the dual SDP (7.35) as the following problem:

$$\text{minimise} \quad \frac{1+r}{d^2} \tag{7.38a}$$

$$\text{subject to} \quad \frac{\rho_{AB} + r\omega}{1+r} = \sigma \otimes \frac{\mathbb{1}}{d}, \tag{7.38b}$$

$$\omega \succcurlyeq 0, \quad \mathrm{tr}(\omega) = 1, \tag{7.38c}$$

$$\sigma \succcurlyeq 0, \quad \mathrm{tr}(\sigma) = 1. \tag{7.38d}$$

The constraints in (7.38) collectively show that this problem—which is no longer written in the form of an SDP—can be interpreted as a generalised robustness. In particular, we can interpret ω as the (generalised) *noise* which must be added to the state ρ_{AB}, in order to bring it to a form where it is a product state and its B system has become maximally mixed. The only small difference is that whereas previously we have been interested in the minimal value of r, now we are interested in the value $\frac{1+r}{d^2}$ instead, which is merely an affine function of r.

What this shows then is that—through duality—we can see that the one-sided optimised singlet fidelity $F_{\Phi^+}^{(A)}(\rho_{AB})$ of a state ρ_{AB}, namely the largest fidelity that can be achieved with a maximally entangled state when optimising over channels on the first system, is in fact equal to the (generalised) robustness of the state to being product and having the second system completely mixed. We hope you will agree that, *a priori*, there seems to be no obvious reason why these two quantities should be related in any way, however through duality, we find the remarkable fact that they are one and the same thing.

The second way to proceed is to return to the dual SDP (7.35) and to simplify it slightly by realising that Z is a slack variable that can be removed. This allows us to re-express (7.35) as

$$\text{minimise} \quad \frac{1}{d}\text{tr}(Y) \tag{7.39a}$$

$$\text{subject to} \quad \rho_{AB} \preccurlyeq Y \otimes \mathbb{1}. \tag{7.39b}$$

In this form, we can now recognise the similarity to the definition of a well-known quantity from quantum information, namely the *conditional min-entropy*. Before jumping straight to this definition, we recall first that the *conditional von Neumann entropy* $S(A|B)_\rho$ can be defined two ways; either directly in terms of the von Neumann entropy as

$$S(A|B)_\rho = S(\rho_{AB}) - S(\rho_B), \tag{7.40}$$

where $S(\rho) = -\text{tr}(\rho \log \rho)$ is the von Neumann entropy; or as a variational expression in terms of the *relative entropy* as

$$S(A|B)_\rho = -\text{minimise} \quad D(\rho_{AB} \| \mathbb{1} \otimes \sigma) \tag{7.41a}$$

$$\text{subject to} \quad \sigma \succcurlyeq 0, \quad \text{tr}(\sigma) = 1, \tag{7.41b}$$

where $D(\rho\|\omega) = \text{tr}(\rho(\log \rho - \log \omega))$ is the quantum relative entropy, and we have used red for the variables, as we are interested in relating to the dual formulation of the optimised singlet fidelity (7.35). It is this second form which generalises to other entropies. In particular, the so-called *min conditional entropy* $H_{\min}(A|B)_\rho$ is defined in terms of the *max relative entropy*[6] $D_{\max}(\rho\|\omega)$, through the analogous equation

$$H_{\min}(A|B)_\rho = -\text{minimise} \quad D_{\max}(\rho_{AB} \| \mathbb{1} \otimes \sigma) \tag{7.42a}$$

$$\text{subject to} \quad \sigma \succcurlyeq 0, \quad \text{tr}(\sigma) = 1, \tag{7.42b}$$

where

$$D_{\max}(\rho\|\omega) = \text{minimise} \quad \lambda \tag{7.43a}$$

$$\text{subject to} \quad \rho \preccurlyeq 2^{\lambda}\omega. \tag{7.43b}$$

In exercise 7.16, it is shown that (7.42) and (7.43) can be combined, in order to give a more direct variational expression for $H_{\min}(A|B)_\rho$, and that furthermore, after introducing new variables,

[6] Note that this terminology—which is standard—might seem confusing, but we follow it here since it does have a justification: the max relative entropy is the largest amongst all Renyi relative entropies, while the associated conditional min-entropy is the smallest amongst all conditional Renyi entropies.

$$H_{\min}(A|B)_{\rho} = -\log \text{minimise} \quad \operatorname{tr}(W) \tag{7.44a}$$

$$\text{subject to} \quad \rho_{AB} \preccurlyeq \mathbb{1} \otimes W \tag{7.44b}$$

$$W \succcurlyeq 0. \tag{7.44c}$$

We are now finally in a position to compare with (7.39). We have arrived at an expression for the conditional min-entropy which is very similar to the expression previously obtained for the dual of the optimised singlet fidelity. The first seemingly important difference is that the constraint is of the form $\rho_{AB} \preccurlyeq \mathbb{1} \otimes W$ rather than $\rho_{AB} \preccurlyeq Y \otimes \mathbb{1}$, however this simply means that the roles of A and B have been interchanged. The other difference is the additional constraint in (7.44) demanding that $W \succcurlyeq 0$, which is not present in (7.35). However, as was argued previously, given that $\rho_{AB} \succcurlyeq 0$, it follows that any feasible W will satisfy $W \succcurlyeq 0$, otherwise (7.44b) cannot be satisfied. We can therefore conclude that the two problems are optimising over the same feasible set (up to the permutation of systems), and therefore we obtain the remarkable result that

$$H_{\min}(B|A)_{\rho} = -\log\left[dF_{\Phi^+}^{(A)}(\rho_{AB})\right]. \tag{7.45}$$

This shows that the conditional min-entropy is a function of the optimised singlet fidelity of a state, and this provides an *operational interpretation* of the quantity.

In the example below, this quantity will be further studied, specialising to a quantum–classical state, and it will be seen that it reduces to a *conditional guessing probability*. Before doing so, we note that the conditional min-entropy—like the conditional (von Neumann) entropy $S(A|B)_{\rho}$—can be *negative*. In fact, from (7.45), the conditional min-entropy will be negative whenever

$$F_{\Phi^+}^{(A)}(\rho_{AB}) > \frac{1}{d}, \tag{7.46}$$

that is, whenever the state ρ_{AB} is sufficiently entangled such that its optimised singlet fidelity is above $\frac{1}{d}$. As we show in exercise 7.17, only entangled states have this property, and hence $H_{\min}(B|A)_{\rho} < 0$ can also be interpreted as an entropic entanglement witness.

Example 7.1. The conditional min-entropy of classical-quantum states.
In this example we will consider a special class of bipartite quantum states, known as 'quantum–classical states', which take the form

$$\rho_{AX} = \sum_x p(x)\rho_A^x \otimes |x\rangle\langle x|, \tag{7.47}$$

where we use X to label the second Hilbert space (rather than B), in order to emphasise that this quantum system encodes a classical random variable X (such that the probability that X takes on value x is $p(x)$), i.e. is diagonal in the basis $\{|x\rangle\}$, and the diagonal entries are the classical probabilities $p(x)$. Such states exhibit purely classical correlations between the classical random variable X, and the quantum states ρ_A^x. In this example, we are interested in computing the conditional min-entropy $H_{\min}(X|A)$ of such states, that is, the conditional min-entropy of the *classical* part of the state, conditioned on the *quantum* part of the state.

Calculating directly using (7.44) is rather complicated in general, so we will instead use (7.45), and thus calculate the optimised singlet fidelity $F_{\Phi^+}^{(A)}(\rho_{AX})$ instead. Going all the way back to (7.32), for a fixed channel $\Lambda(\cdot)$, we see that

$$\langle\Phi^+|\Lambda \otimes \mathrm{id}(\rho_{AX})|\Phi^+\rangle = \langle\Phi^+|\left(\sum_x p(x)\Lambda(\rho_A^x) \otimes |x\rangle\langle x|\right)|\Phi^+\rangle, \tag{7.48a}$$

$$= \frac{1}{d}\sum_i \langle i|\langle i|\left(\sum_x p(x)\Lambda(\rho_A^x) \otimes |x\rangle\langle x|\right)\sum_j |j\rangle|j\rangle, \tag{7.48b}$$

$$= \frac{1}{d}\sum_x p(x)\langle x|\Lambda(\rho_A^x)|x\rangle, \tag{7.48c}$$

$$= \frac{1}{d}\sum_x p(x)\mathrm{tr}[\Lambda(\rho_A^x)|x\rangle\langle x|], \tag{7.48d}$$

$$= \frac{1}{d}\sum_x p(x)\mathrm{tr}[\rho_A^x\Lambda^\dagger(|x\rangle\langle x|)]. \tag{7.48e}$$

In the above, in the first line we substituted the definition of the quantum–classical state; in the second line we substituted for the definition of $|\Phi^+\rangle$; and in the third we performed the inner products and simplified. Finally, in the fourth and fifth line we first expressed the expectation value as a trace, and then used the definition of the adjoint map to move the action of $\Lambda(\cdot)$ onto the projector $|x\rangle\langle x|$.

In the final line, an important realisation can be made. Since $\Lambda^\dagger(\cdot)$ is a completely positive *unital map*, it maps measurements to measurements. That is, given any measurement $\mathbb{M} = \{M_x\}$, a new measurement $\mathbb{M}' = \{M_x'\}$ with POVM elements $M_x' = \Lambda^\dagger(M_x)$ we can formed, using any completely positive unital map $\Lambda^\dagger(\cdot)$. Moreover, as shown in exercise 7.18 *any d* outcome measurement acting on a *d*-dimensional Hilbert space can be formed by applying a completely positive unital map to an ideal projective measurement of the form $\mathbb{M} = \{|x\rangle\langle x|\}$.

When considering optimising over all channels $\Lambda(\cdot)$, we can now use the above realisation to re-express this as *an optimisation over all d-outcome measurements $\mathbb{M}$*. That is, for quantum–classical states, an equivalent formulation of the optimised singlet fidelity is

$$F_{\Phi^+}^{(A)}(\rho_{AX}) = \text{maximise} \quad \frac{1}{d}\sum_x p(x)\mathrm{tr}[\rho_A^x M_x] \tag{7.49a}$$

$$\text{subject to} \quad M_x \succcurlyeq 0, \quad \sum_x M_x = \mathbb{1}. \tag{7.49b}$$

This is however exactly the same problem (up to the factor of $\frac{1}{d}$) that was encountered previously in chapter 4, namely this is *quantum state discrimination*, as given in (4.9).

That is to say that we can view the quantum–classical state (7.47) as encoding a quantum state discrimination task, and see that the optimised singlet fidelity of this state is nothing but $\frac{1}{d}P^*_{\text{guess}}$, i.e. it is up to the factor of $\frac{1}{d}$ the guessing probability in quantum state discrimination that was previously studied.

Combining this with (7.45), we obtain the important result that for quantum–classical states,

$$H_{\min}(X|A)_\rho = -\log P^*_{\text{guess}}, \tag{7.50}$$

that is, the conditional min-entropy is precisely the log of the guessing probability. We can view the quantum states ρ_A^x as *quantum side information* about the classical variable X. The conditional min-entropy is the (conditional) guessing probability of correctly identifying the value of the random variable X. This shows that the optimised singlet fidelity $F^{(A)}_{\Phi^+}(\rho_{AB})$ can be viewed as a *fully quantum* generalisation of this guessing task. That is, the generalisation of guessing the value of a random variable can be thought of as maximising the (quantum) correlation—i.e. entanglement—between the system

Exercises

7.14

(a) Show that the maximally entangled state $|\Phi^+\rangle = \frac{1}{\sqrt{d}}\sum_i |i\rangle|i\rangle$ has the following invariance property:

$$(A \otimes \mathbb{1})|\Phi^+\rangle = (\mathbb{1} \otimes A^\mathsf{T})|\Phi^+\rangle, \tag{7.51}$$

for any operator A, where $^\mathsf{T}$ denotes transpose in the basis of $|\Phi^+\rangle$. *Hint: One way to show this is to consider the matrix elements of the left- and right-hand side, and so show that they are equal.*

(b) Using (7.51), show that all of the states that $|\Phi^+\rangle$ can be transformed into by local unitaries on both systems, i.e. all states of the form $(U \otimes V)|\Phi^+\rangle$, can be reached by considering unitaries on the first system alone, i.e. by transformations of the form $(W \otimes \mathbb{1})|\Phi^+\rangle$, for W a unitary operator.

7.15 In this exercise we will derive the dual formulation of (7.34), as given in (7.35).

(a) Write down the Lagrangian associated with (7.34), calling the dual variables corresponding to the first and second constraint respectively Y and Z.

(b) Use the Lagrangian to derive the dual formulation (7.35).

(c) Show that both the primal (7.34) and dual (7.35) are strictly feasible, and hence that strong duality holds.

7.16 In this exercise we will show how (7.42) and (7.43) can be combined, in order to give a more direct variational expression for $H_{\min}(A|B)_\rho$ as given in (7.44).

(a) Show that by combining (7.42) and (7.43) directly, one obtains the following optimisation problem for $H_{\min}(A|B)_\rho$:

$$H_{\min}(A|B)_\rho = -\text{minimise} \quad \lambda \tag{7.52a}$$

$$\text{subject to} \quad \rho_{AB} \preccurlyeq 2^\lambda \mathbb{1} \otimes \sigma, \tag{7.52b}$$

$$\sigma \succcurlyeq 0, \quad \text{tr}(\sigma) = 1, \tag{7.52c}$$

This is not an SDP, since (i) the variable λ appears in the exponent of the first constraint; (ii) The variables λ and σ appear multiplied together.

(b) Introduce the new variable $W = 2^\lambda\sigma$. Show that (7.52) can be re-expressed solely in terms of W, as

$$H_{\min}(A|B)_\rho = -\text{minimise} \quad \log[\text{tr}(W)] \tag{7.53a}$$

$$\text{subject to} \quad \rho_{AB} \preccurlyeq \mathbb{1} \otimes W, \tag{7.53b}$$

$$W \succcurlyeq 0. \tag{7.53c}$$

(c) Explain why $\min_x \log(x) = \log(\min_x x)$, and use this to show that (7.53) is equivalent to (7.44).

7.17 In this exercise we will show that the optimised singlet fidelity is less than $\frac{1}{d}$ for all separable states, $F_{\Phi^+}^{(A)}(\sigma_{\text{sep}}) \leqslant \frac{1}{d}$.

(a) Consider first product states of the form $\sigma_{\text{sep}} = \omega \otimes \zeta$. Show that in this case, the optimised singlet fidelity is:

$$F_{\Phi^+}^{(A)}(\omega \otimes \zeta) = \text{maximise} \quad \langle\Phi^+|\sigma \otimes \zeta|\Phi^+\rangle \tag{7.54a}$$

$$\text{subject to} \quad \sigma \succcurlyeq 0, \quad \text{tr}(\sigma) = 1. \tag{7.54b}$$

(b) Show that for all operators F and G that

$$\langle\Phi^+|F \otimes G|\Phi^+\rangle = \frac{1}{d}\text{tr}[F^\top G]. \tag{7.55}$$

(c) Use part (a) and (b) to show that $F_{\Phi^+}^{(A)}(\omega \otimes \zeta) \leqslant \frac{1}{d}$.

(d) Show that the optimised singlet fidelity is a *convex* function, that is

$$F_{\Phi^+}^{(A)}\left(\sum_i p(i)\sigma_i\right) \leqslant \sum_i p(i) F_{\Phi^+}^{(A)}(\sigma_i), \tag{7.56}$$

for any ensemble of states $\{p(i), \sigma_i\}$.

(e) Use part (c) and (d) to show that the optimised singlet fidelity is less than $\frac{1}{d}$ for all separable states.

7.18 In this exercise we will show that *any* d outcome measurement acting on a d-dimensional Hilbert space can be formed by applying a completely positive unital map to an ideal projective measurement of the form $\mathbb{M} = \{|x\rangle\langle x|\}$. Consider the following Choi state

$$X_{\Lambda^\dagger} = \frac{1}{d}\sum_a M_a \otimes |a\rangle\langle a|, \tag{7.57}$$

where $\mathbb{M} = \{M_a\}$ is an arbitrary d outcome POVM.

(a) Show that $X_{\Lambda^\dagger} \succcurlyeq 0$.

(b) Show that $\text{tr}_B(X_{\Lambda^\dagger}) = \frac{1}{d}\mathbb{1}$.

This shows that $X_{\Lambda^\dagger}$ is the Choi state associated to a completely positive unital map $\Lambda^\dagger(\cdot)$.

(c) Show that the action of the associated channel $\Lambda^\dagger(\cdot)$ on the projector $|x\rangle\langle x|$ is

$$\Lambda^{\dagger}(|x\rangle\langle x|) = M_x. \tag{7.58}$$

This shows that it is possible to map an ideal projective measurement (acting on a d-dimensional Hilbert space) into any other d outcome measurement $\mathbb{M}$.

7.5 Concluding remarks

In the last chapter of this book we have dealt with quantum channels, the objects that describes how quantum states evolve or can be manipulated. The key fact that allows the use of SDPs in the study of quantum channels is the Choi–Jamiołkowski isomorphism between channels and states. In particular, we presented the following topics:

- **Channel estimation.** Given a set of input and output states of an unknown channel, it is possible to determined which channels are consistent with this data using an SDP—see equation (7.15).
- **Diamond norm.** We can compute the diamond norm of a channel and the diamond norm distance between channels using semidefinite programming—see equation (7.23).
- **Channel discrimination.** The average success probability in a binary channel discrimination task (7.20) can be calculated by an SDP, and provides an operational interpretation of the diamond norm distance—see (7.22).
- **Optimised singlet fidelity and generalised robustness.** The optimised singlet fidelity quantifies how close we can bring a quantum state to a maximally entangled state by applying local quantum channels on one of the subsystems. It can be calculated using the SDP (7.34). Furthermore, its dual formulation, as given in equation (7.35) provides us with an operational interpretation of the optimised singlet fidelity as a generalised robustness—see equation (7.38).
- **Optimised singlet fidelity and conditional min-entropy.** Duality furthermore provides us with a one-to-one connection between the conditional min-entropy of a bipartite state (equations (7.40) and (7.41)) and the optimised singlet fidelity of that state—see equation (7.45).

7.6 Further reading

- Watrous J 2018 *Theory of Quantum Information* (Lecture Notes) (Cambridge: Cambridge University Press) https://cs.uwaterloo.ca/~watrous/TQI/
- https://cs.uwaterloo.ca/~watrous/TQI-notes/

7.7 Advanced topics

In this brief section we will show how to arrive at the SDP (7.23) for the diamond norm. Our starting point will be (7.21), which we repeat for convenience:

$$\|\Gamma(\cdot)\|_\diamond = \text{maximise} \quad \text{tr}[Z(\Gamma \otimes \text{id}(\rho))] \tag{7.59a}$$

$$\text{subject to} \quad -\mathbb{1} \preccurlyeq Z \preccurlyeq \mathbb{1}, \tag{7.59b}$$

$$\rho \succcurlyeq 0, \quad \text{tr}(\rho) = 1. \tag{7.59c}$$

This form is not an SDP, since the variables Z and ρ appear multiplied together in the objective function. In this section, we will see how to combine these variables, which will lead us to the SDP formulation.

The first step we will take is to reformulate the problem in terms of the Choi state X_Γ associated to the map $\Gamma(\cdot)$.[7] Now, (7.4) gives the action of a map $\Gamma(\cdot)$ on a state ρ, in terms of its Choi state. However, this doesn't specify the action of $\Gamma \otimes \text{id}$ on a bipartite state ρ_{AB}. We can however extend (7.4) linearly, and see that the action is

$$\Gamma \otimes \text{id}(\rho_{AC}) = d\,\text{tr}_B\left[(\mathbb{1}_A \otimes \rho_{BC}^{\mathsf{T}_B})(X_{AB} \otimes \mathbb{1}_C)\right], \tag{7.60}$$

where we have introduced labels for all of the Hilbert spaces, in order to make everything as clear as possible. In particular, note that it is convenient to treat the state as being on the Hilbert spaces A and C, in order to allow the Hilbert space B to be that of the Choi state.

Using (7.60), and introducing explicit labels for the Hilbert spaces, we see that the objective function in (7.59) can be written as

$$\text{tr}[Z_{AC}(\Gamma \otimes \text{id}(\rho_{AC}))] = d\,\text{tr}[(Z_{AC} \otimes \mathbb{1}_B)(\mathbb{1}_A \otimes \rho_{BC}^{\mathsf{T}_B})(X_{AB} \otimes \mathbb{1}_C)]. \tag{7.61}$$

Note that in the above, there is a slight complication with regards to the notation. In particular, the first term on the right-hand side, $(Z_{AC} \otimes \mathbb{1}_B)$ does *not* mean that the ordering of the Hilbert spaces B and C has been interchanged. We still want to consider that the order of the Hilbert spaces is the (standard) order ABC. This is however difficult to denote in writing. This is why we choose to take the approach of labelling the Hilbert spaces, and accept that the order in which the Hilbert spaces are written down may disagree with the actual order in which the Hilbert spaces occur.

With the above subtlety aside, we can write the right-hand side of (7.61) in the form $\text{tr}(Y_{AB}X_{AB})$ if the following operator is defined

$$Y_{AB} = d\,\text{tr}_C[(Z_{AC} \otimes \mathbb{1}_B)(\mathbb{1}_A \otimes \rho_{BC}^{\mathsf{T}_B})], \tag{7.62}$$

which combines together the two variables Z and ρ into a single operator variable. Our goal is thus to characterise the constraints that are satisfied by Y, given the constraints that are satisfied by Z and ρ from (7.59b) and (7.59c), and to show that they are linear equality and/or inequality constraints, which can then be imposed inside an SDP.

[7] We remind, once again, that since $\Gamma(\cdot)$ is not necessarily either completely positive or trace preserving, strictly speaking X_Γ is not a *state*, but we can think of it as a *pseudo-state*, since it still arises from the local action of the map $\Gamma(\cdot)$ on the state $|\Phi^+\rangle$.

The first step we will take is to realise that the partial transpose can be removed by introducing a maximally entangled state and performing a partial trace. In particular, in exercise 7.20 it is shown that

$$\mathrm{tr}_A[(F \otimes \mathbb{I})|\Phi^+\rangle\langle\Phi^+|] = \frac{1}{d}F^{\mathsf{T}}, \tag{7.63}$$

for any operator F. Using this, we can express Y as

$$Y_{AB} = d^2\mathrm{tr}_{CD}\left[(Z_{AC} \otimes |\Phi^+\rangle\langle\Phi^+|_{BD})(\mathbb{I}_{AB} \otimes \rho_{DC})\right]. \tag{7.64}$$

Now, based upon this expression, we can consider the following class of maps

$$\Omega_\omega(G) = d^2\mathrm{tr}_{CD}[(G_{AC} \otimes |\Phi^+\rangle\langle\Phi^+|_{BD})(\mathbb{I}_{AB} \otimes \omega_{CD})], \tag{7.65}$$

where $\omega \succcurlyeq 0$ is a positive semidefinite operator. It is straightforward to see that this map is positive, such that $\Omega_\omega(G) \succcurlyeq 0$ whenever $G \succcurlyeq 0$. In particular, we can see that

$$\mathrm{tr}[F\Omega_\omega(G)] = d^2\mathrm{tr}[(G_{AC} \otimes |\Phi^+\rangle\langle\Phi^+|_{BD})(F_{AB} \otimes \omega_{CD})] \succcurlyeq 0, \tag{7.66}$$

whenever $F \succcurlyeq 0$, since in this case all operators on the right-hand side are positive semidefinite. Thus this shows that $\Omega_\omega(G) \succcurlyeq 0$. The reason for introducing this map is because we can then see that $Y = \Omega_\rho(Z)$. This can be used in the following way: from the left-hand constraint in (7.59b), we see that $\mathbb{I} + Z \succcurlyeq 0$. From the fact that $\Omega_\rho(\cdot)$ is a positive map, we can therefore conclude that $\Omega_\rho(\mathbb{I} + Z) \succcurlyeq 0$. Since Ω_ρ is also linear,

$$\Omega_\rho(\mathbb{I} + Z) = \Omega_\rho(\mathbb{I}) + \Omega_\rho(Z) \tag{7.67a}$$

$$= \Omega_\rho(\mathbb{I}) + Y, \tag{7.67b}$$

$$= d\mathbb{I} \otimes \rho_B^{\mathsf{T}} + Y, \tag{7.67c}$$

where in the last line we used

$$\Omega_\rho(\mathbb{I}) = d^2\mathrm{tr}_{CD}\left[(\mathbb{I}_{AC} \otimes |\Phi^+\rangle\langle\Phi^+|_{BD})(\mathbb{I}_{AB} \otimes \rho_{CD})\right], \tag{7.68a}$$

$$= d^2\mathbb{I}_A \otimes \mathrm{tr}_D[|\Phi^+\rangle\langle\Phi^+|_{BD}(\mathbb{I}_B \otimes \rho_D)], \tag{7.68b}$$

$$= d(\mathbb{I} \otimes \rho_B^{\mathsf{T}}). \tag{7.68c}$$

Thus, if we denote by

$$\sigma = \rho_B^{\mathsf{T}}, \tag{7.69}$$

we conclude that $\Omega_\rho(\mathbb{I} + Z) \succcurlyeq 0$ implies that

$$-d(\mathbb{I} \otimes \sigma) \preccurlyeq Y. \tag{7.70}$$

An identical line of reasoning follows for the right-hand constraint in (7.59b), from which we arrive at

$$Y \preccurlyeq d(\mathbb{1} \otimes \sigma). \tag{7.71}$$

We have thus arrived at a new pair of variables, (Y, σ), which are constrained to satisfy $-\mathbb{1} \otimes \sigma \preccurlyeq Y \preccurlyeq \mathbb{1} \otimes \sigma$ and $\sigma \succcurlyeq 0$, $\mathrm{tr}(\sigma) = 1$, where the latter two constraints follow since σ is the transpose of the reduced density operator of ρ, which is itself a density operator. What will now be shown is that from any such pair of operators we can construct Z and ρ which satisfy the constraints (7.59b) and (7.59c), which shows that the two pairs are equivalent.

Concerning ρ, we will take this to be any extension of σ^{T}. That is, we can take ρ to be any bipartite density operator such that $\mathrm{tr}_A(\rho) = \sigma_B^{\mathsf{T}}$. Being a valid density operator, this means that $\mathrm{tr}(\rho) = 1$ and $\rho \succcurlyeq 0$, as required by (7.59c).

For Z we will take the operator

$$Z = \frac{1}{d}\left(\mathbb{1} \otimes \frac{1}{\sqrt{\sigma}}\right) Y \left(\mathbb{1} \otimes \frac{1}{\sqrt{\sigma}}\right). \tag{7.72}$$

Here, as previously, we take $\frac{1}{\sqrt{\sigma}}$ to be defined by the pseudo-inverse, i.e. only upon its support (and to remain zero outside of its support). The map $\Gamma_\sigma(F) = \frac{1}{d}\left(\mathbb{1} \otimes \frac{1}{\sqrt{\sigma}}\right) F \left(\mathbb{1} \otimes \frac{1}{\sqrt{\sigma}}\right)$, being a conjugation, is positive. We can therefore apply it to $Y + d(\mathbb{1} \otimes \sigma) \succcurlyeq 0$, and will have $\Gamma_\sigma(Y + d(\mathbb{1} \otimes \sigma)) \succcurlyeq 0$. This implies, in particular that

$$\Gamma_\sigma(Y + d(\mathbb{1} \otimes \sigma)) = \frac{1}{d}\left(\mathbb{1} \otimes \frac{1}{\sqrt{\sigma}}\right)(Y + d(\mathbb{1} \otimes \sigma))\left(\mathbb{1} \otimes \frac{1}{\sqrt{\sigma}}\right), \tag{7.73a}$$

$$= Z + \mathbb{1} \otimes \mathbb{1} \succcurlyeq 0. \tag{7.73b}$$

Similarly, since $d(\mathbb{1} \otimes \sigma) - Y \succcurlyeq 0$, we have $\Gamma_\sigma(d(\mathbb{1} \otimes \sigma) - \sigma) \succcurlyeq 0$, which an identical calculation shows that

$$\mathbb{1} \otimes \mathbb{1} - Z \succcurlyeq 0. \tag{7.74}$$

Thus, we have $-\mathbb{1} \preccurlyeq Z \preccurlyeq \mathbb{1}$, exactly the constraints (7.59b).

To summarise, we have shown that on the one hand, for any pair of variables (Z, ρ) from the set

$$\{(Z, \rho) \mid -\mathbb{1} \preccurlyeq Z \preccurlyeq \mathbb{1}, \rho \succcurlyeq 0, \mathrm{tr}(\rho) = 1\} \tag{7.75}$$

we can find a pair of variables (Y, σ) from the set

$$\{(Y, \sigma) \mid -d(\mathbb{1} \otimes \sigma) \preccurlyeq Y \preccurlyeq d(\mathbb{1} \otimes \sigma), \sigma \succcurlyeq 0, \mathrm{tr}(\sigma) = 1\}, \tag{7.76}$$

through the mappings (7.62) and (7.69). On the other hand, for any pair (Y, σ) we can find a pair (Z, ρ) using the inverse mapping (7.72), and taking ρ to be any extension of σ^{T}. Finally, we saw that the bi-linear objective function (7.61) of ρ and Z is equal to the linear objective function of Y, $\mathrm{tr}(YX)$. Altogether, this therefore shows that the diamond norm can be written as an SDP in the variables (Y, σ) as given in (7.23), namely

$$\|\Gamma(\cdot)\|_{\diamond} = \text{maximise} \quad \mathrm{tr}[YX_{\Gamma}] \tag{7.77a}$$

$$\text{subject to} \quad -d(\mathbb{1} \otimes \sigma) \preccurlyeq Y \preccurlyeq d(\mathbb{1} \otimes \sigma), \tag{7.77b}$$

$$\sigma \succcurlyeq 0, \quad \mathrm{tr}(\sigma) = 1. \tag{7.77c}$$

Exercises

7.19 Show that in (7.60), if we consider a product state of the form $\rho_{AC} = \rho_A \otimes \rho_C$, then this expression recovers (7.4), up to the addition of the system ρ_C.

7.20 Show that the maximally entangled state satisfies the property (7.63).

www.ingramcontent.com/pod-product-compliance
Ingram Content Group UK Ltd.
Pitfield, Milton Keynes, MK11 3LW, UK
UKHW051852100926
8556IPUK00005B/587